T0094362

Physicochemical Behavior and Supramolecular Organization of Polymers

Ligia Gargallo · Deodato Radić

Physicochemical Behavior and Supramolecular Organization of Polymers

 Springer

Prof. Dr. Ligia Gargallo
Pontificia Universidad Catolica
de Chile
Facultad de Quimica
Vicuna Mackenna 4860
Santiago
Casilla 306, Correo 22
Chile
lgargall@puc.cl

Prof. Dr. Deodato Radić
Pontificia Universidad Catolica
de Chile
Facultad de Quimica
Vicuna Mackenna 4860
Santiago
Casilla 306, Correo 22
Chile
dradic@uc.cl

ISBN 978-1-4020-9371-5   e-ISBN 978-1-4020-9372-2

DOI 10.1007/978-1-4020-9372-2

Library of Congress Control Number: 2008942790

Printed on acid-free paper

9 8 7 6 5 4 3 2 1

springer.com

Preface

This book is concerned mainly with the physicochemical behavior and supramolecular organization of polymers. The book is split in four chapters dealing with solution properties, viscoelastic behavior, physicochemical aspects at interfaces and supramolecular structures of polymeric systems. The classical treatment of the physicochemical behavior of polymers is presented in such a way that the book will meet the requirements of a beginner in the study of polymeric systems in solution and in some aspects of the solid state, as well as those of the experienced worker in other type of material. Indeed the book is a contribution to the chemistry of materials. Taken into account these aspects, Chapter 1 is an introduction to the classical conformational and thermodynamic analysis of polymeric solutions where the different theories that describe these behaviors of polymers are analyzed. Owing to the importance of the basic knowledge of the solution properties of polymers, the description of the conformational and thermodynamic behavior of polymers is presented in a classical way. The basic concepts like theta condition, excluded volume, good and poor solvents, critical phenomena, concentration regime, cosolvent effect of polymers in binary solvents, preferential adsorption are presented in an intelligible way. The thermodynamic theory of association equilibria which is capable to describe quantitatively the preferential adsorption of polymers by polar binary solvents is also analyzed. Chapter 2 is a discussion of the viscoelastic properties of polymeric material where the different concept dealing with the fact that polymers above glass-transition temperature exhibit high entropic elasticity. Polymers exhibit both viscous and elastic characteristics what is present in systems when undergoing deformation. In this Chapter the basic concepts of viscoelasticity are described at beginner level. The analysis of stress-strain in polymeric materials is of great practical interest and several examples of some familiar behavior of polymeric materials are shortly described. The Chapter is splitted in four parts the first dealing with basic concepts of viscoelasticity. The second with dielectric and dynamic mechanical behavior of aliphatic, cyclic saturated and aromatic substituted poly(methacrylate)s with different kind of substituents in the side rings. The discussion in terms of the theories that can describe the viscoelastic behavior of polymers is well explained. The analysis of the different relaxations that take place in these systems allow to understand the molecular origin of the different motions. By this way an interesting approach of the relaxational processes is presented under the experience of the

authors in these polymeric systems. The third part deals with the dielectric and
dynamic mechanical behavior of poly(itaconate)s with mono and disubstitutions.
The effect of the substituents and the free carboxylic groups in poly(monitaconate)s
and the disubstitution on poly(diitaconate)s is extensively discussed and interesting
conclusion are descrived. The fourth part is the analysis of viscolastic behavior of
poly(thiocarbonate)s where the difference is that this family of polymers correspond
to condensation polymers instead of vinyl polymers like the formers. The effect of
the substitution of the polymers is also analyzed. Chapter 3 is a discussion of the
behavior of polymers at interfaces where the Langmuir monolayers and Langmuir-
Blodget films are studied. Amphiphilic polymers at the air-water interface are studied via the Langmuir technique. The study and discussion of surface pressure-area
isotherms for different polymers are performed by using a surface film balance and
the results obtained from this technique are analyzed in terms of the shape of the
isotherms. The collapse pressure for different systems are discussed in terms of the
chemical structure of the polymer. The adsorption of polymers by spreading and
from solution is also discussed. Wetting of solids by a liquid described in terms of
the equilibrium contact angle θ and the appropriate interfacial tensions. At equilibrium the forces acting are analyzed using the Young's equation. Chapter 4 deals
with the analysis of supramolecular structures containing polymers. Specifically in
this chapter the discussion about the effect of polymeric materials with different
chemical structures that form inclusion complexes is extensively studied. The effect of the inclusion complexes at the air-water interface is discussed in terms on
the nature of the interaction i.e. if the interaction is on entropic or enthalpic nature. The description of these inclusion complexes on different cyclodextrines with
poly(ethylene) oxide, poly(ε-caprolactone) and related polymers is an interesting
way to understand some non-covalent interaction in these systems. The discussion
about the generation and effect of supramolecular structures on molecular assembly
and auto-organization processes is also presented in a single form. Finally the use of
block copolymers and dendronized polymers at interfaces is new aspect to be taken
into account from both basic and technological interest. The effect of the chemical
structure on the self-assembled systems is discussed in terms of the different kinds
of interaction that can be detected. This book should be a powerful tool for students
and scientists working both in polymer chemistry and physic and in material science.

Santiago, Chile   Ligia Gargallo
                                                                                   Deodato Radić

Contents

Abbreviations

a	activity
a.c	alternating current
α	thermal expansion coefficient
α	relaxation associated to T_g
$A_i, i = 2, 3, \ldots$	ith virial coefficient
A	area
Å_o	surface limiting area
Å_c	surface critical area
AFM	atomic force microscopy
BAM	Brewster Angle Microscopy
C	concentration
c^*	concentration at which entanglements set in
c^*	overlap concentration
C_k	Kuhmian concentration
C_s	two dimensional compressibility
C_∞	characteristic ratio
CMC	critical micelle concentration
CD	cyclodextrin
D	docility
D	diffusion coefficient
Φ	density
Φ'	universal Flory constant
DLS	dynamic light scattering
DMA	dynamic mechanical analysis
DMF	N,N-dimethylformamide
DMSO	dimethylsulfoxide
ρ	density
DSC	differential scanning calorimeter
$d\varepsilon/dw$	parameter proportional to the total polarization of the chains
δ	solubility parameter of Hildebrand
δ	the lag in the phase angle. G:dynamic modulus
$d\varepsilon/dt$	the time derivative of strain.
ε	static comprensibility modulus

ε	strain that occur under the given stress
ε^*	complex permittivity
ε'	dielectric permittivity
ε_1	dielectric permittivity
ε''	dielectric loss
ε_o	relaxed permittivity
ε_∞	unrelaxed permittivity
E_a	activation energy
E	electrostatic modulus (Young's modulus)
ε	strain that occur under the given stress.
E'	real part of viscoelastic spectra
E''	imaginary part of viscoelastic spectra.
$E^{\mp}$	the apparent activation energy
Eg	glassy modulus
ΔE	internal energy
ΔE_V	change in the molar internal energy
g	phenomenological interaction parameter
g_T	ternary parameter
G	free energy
G	generation
G'	dynamic stress modulus, storage modulus
G''	dynamic strain modulus, loss modulus.
ΔG	change in free energy
ΔG_M	change in free energy by mixing
ΔG^0	change in standard free energy
GPC	gel permeation chromatography (SEC)
f	fugacity
H	heat
ΔH	change in enthalpy
ΔH_V	change in the molar enthalpy
F(t)	relaxation function
F_{max}	frequency at the maximum of the isotherm
f_{ax}	axial conformation
f_{eq}	equatorial conformation.
F	strength
FEM	transmission electron microscopy
IC	inclusion complex
I	second momento of area of the cross section.
I	intensity of scattering radiation
I_0	intensity of incident radiation
IR	infrared
J	compliance
h	Plank's constant
k'	Huggins viscosity constant
K	Mark-Houwink pre-exponent coefficient

K_i	equilibrium constants
K_θ	Mark-Houwink coefficient at θ conditions
$K(t)$	creep
ΔL	difference in the length
l	Debye length
LB	Langmuir-Blodgett
LCST	lower critical solution temperature
m'	parameter of the VFTH equation
m	parameter dealing with the broadness of a relaxation
MMX	force field
MM2P	force field
MD	molecular dynamic
MKS	Mark-Houwink-Sakurada
MDS	molecular Dynamic Simulation
M	mass, molecular weight
M^*	modulus
M_0	molecular weight of polymer repeating unit
M_n	number-average molecular weight
M_v	viscosity-average molecular weight
M_w	weight-average molecular weight
n	critical exponent of the excluded volume
$\Delta\mu_1$	change in chemical potential
n	refractive index of solution
n_0	refractive index of solvent
dn/dc	limiting value of the specific refractive index increment at zero concentration
N	Avogadro's number
N	crosslink density
HN	Havriliak-Negami
NMR	nuclear magnetic resonance
P2CEM	poly(2-chloroethyl methacrylate)
P3CEM	poly(3-propyl methacrylate)
P2ClCHA	poly(2-chlorocyclohexyl acrylate)
PCHM	poly(cyclohexyl methacrylate)
PCHMM	poly(cyclohexylmethyl methacrylate)
PCHPM	poly(cyclohexylpropyl methacrylate)
PCHBM	poly(cyclohexylbutyl methacrylate)
P2CEM	poly(2-chloroethyl methacrylate)
P2tMCHM	poly(2-tert-butycyclohexyl methacrylate)
P4tMCHM	poly(4-tert-buycyclohexyl methacrylate)
PCHpM	poly(cyclohepthyl methacrylate)
PCHpMM	poly(cycloheptylmethy methacrylate)
PCOcM	poly(cyclooctyl methacrylate)
PCBuM	poly(cyclobutyl methacrylate)
PCBMM	poly(cyclobutylmethyl methacrylate)

P2NBM	poly(2-norbornyl methacrylate)
P3M2NBM	poly(3-methyl-2-norbornyl methacrylate)
P4THPMA	poly(tetrahydropyranyl methacrylate)
PDMA	poly(1,3-dioxan-5-yl-methacrylate)
PTHFM	poly(tetrahydrofurfuryl methacrylate)
P3MTHFMA	poly(3-methyl-tetrahydrofurfuryl methacrylate)
PPHM	poly(phenyl methacrylate)
P2,6DMPM	poly(2,6-dimethylphenyl methacrylate)
P2,4DMPM	poly(2,4-dimethylphenyl methacrylate)
P2,5DMPM	poly(2,5-dimethylphenyl methacrylate)
P3,5DMPM	poly(3,5-dimethylphenyl methacrylate)
P2,4DFBM	poly(2,4-difluorobenzyl methacrylate)
P2,5DFBM	poly(2,5-difluorobenzyl methacrylate)
P2,6DFBM	poly(2,6-difluorobenzyl methacrylate)
P2MClBM	poly(2-monochlorobenzyl methacrylate)
P3MClBM	poly(3-monochlorobenzyl methacrylate)
P4MClBM	poly(4-monochlorobenzyl methacrylate)
P2,3DClBM	poly(2,3-dichlorobenzyl methacrylate)
P2,4DClBM	poly(2,4-dichlorobenzyl methacrylate)
P2,5DClBM	poly(2,5-dichlorobenzyl methacrylate)
P2,6DClBM	poly(2,6-dichlorobenzyl methacrylate)
P3,4DClBM	poly(3,4-dichlorobenzyl methacrylate)
P3,5DClBM	poly(3,5 -dichlorobenzyl methacrylate)
PMOI	poly(monooctyl itaconate)
PMDI	poly(monodecyl itaconate)
PDMI	poly(dimethyl itaconate)
PDEI	poly(diethyl itaconate)
PDPI	poly(dipropyl itaconate)
PDBI	poly(dibutyl itaconate)
PDIPI	poly(diisopropyl itaconate)
PDIBI	poly(diisobutyl itaconate)
PMMA	poly(methyl methacrylate)
PMCHI	poly(monocyclohexyl itaconate)
PDCHI	poly(dicyclohexyl itaconate)
PDCHpI	poly(dicycloheptyl itaconate)
PDCOcI	poly(dicyclooctyl itaconate)
PDCBI	poly(dicyclobutylitaconate)
POS	poly(octamethylene sebacamide)
POT	poly(octamethylene terephthalamide)
POTCl	poly(octamethylene tetrachloroterephthalamide)
PTC	poly(thiocarbonate)
PVP	poly(N-vinil-2-pyrrolidone)
PS	poly(styrene)
PIB	poly(isobutylene)
PEC	poly(ε-caprolactone)

PEO	poly (ethylene oxide)
Q	first moment of area
r_o^2	unperturbed mean square dimension.
r_{of}^2	free rotation unperurbed mean square dimension
R	gas constant
R_H	hydrodynamic radius
R_η	hydrodynamic radius
R_g	radius of gyration
σ	rigidity factor
σ	applied stress, shear strain. sinusoidal stress response
SDS	sodium dodecyl sulfate
STM	scanning tunneling microscopy
SANS	small- angle neutron scattering
π_C	critical surface pressure
S, ΔS	entropy, change in entropy
ΔS^*_M	change in configurational entropy
ΔS^E_M	change in entropy by mixing
SEC	size-exclusion chromatofraphy (GPC)
$<S^2>$	mean-square radius of gyration
η_{coil}	average density of segments
T_g	glass transition temperature
t	time
T_{max}	temperature where E" has the maximum value.
T_∞	parameter of the VFTH equation
T_0	intial temperature
Γ	surface concentration
T_f	final temperature
T_p	polarization temperature
T_a	anneal temperature
$\tau(T)$	relaxation time related with the depolarization current iT
THF	tetrahydrofuran
τ	shear stress
Θ	theta temperature
τ	shear stress
t	thickness in the material perpendicular to the shear
i.T	depolarization current
tanδ:	$G''/G' = \gamma/\sigma$
UCST	upper critical solution temperature
V	molar volume
Vsp	specific volume
V	shear force
ν	rate of conformational change
v	Mark-Houwink power coefficient
v_i	volume fraction
VFTH	Vogel, Fulcher, Tamman, Hesse equation

ω	frequency
dW	work (fdX)
Wd	adhesion work
Ψ	entropic contribution to χ
γ	monolayer surface tension
γ°	water surface tension
χ_{crit}	critical interaction parameter
χ	phenomenological interaction parameter for noncombinatorial part
χ	phenomenological interaction parameter
λ	preferential adsorption coefficient
ξ	screening length
Ø	segment fraction
π	surface pressure
π	osmotic pressure
π/c	reduced osmotic pressure
η_S	surface viscosity
η	solution or melt viscosity
η_0	viscosity at zero shear rate
η_{sp}	specific viscosity
[η]	intrinsic viscosity
γ	surface tension
$\gamma_{S/V}$	interfacial tension at the solid/vapour
$\gamma_{S/L}$	interfacial tension at the solid/liquid
$\gamma_{L/V}$	interfacial tension at the liquid/vapour
γ	sinusoidal oscillatory shear strain
γ	surface tension
γ	activity coefficient
γ_0	initial sinusoidal oscillatory strain
λ	extension ratio $= L/L_0$
γ	subglass relaxation
β	subglass relaxation
δ	subglass relaxation
$\frac{\phi}{B}$	free volume
$\frac{\langle \mu^2 \rangle}{x}$	mean square dipole moment per polymer repeating unit
π/c	reduced osmotic pressure

Chapter 1
Polymer Solution Behavior: Polymer in Pure Solvent and in Mixed Solvent

Summary The classical treatment of the physicochemical behavior of polymers is presented in such a way that the chapter will meet the requirements of a beginner in the study of polymeric systems in solution. This chapter is an introduction to the classical conformational and thermodynamic analysis of polymeric solutions where the different theories that describe these behaviors of polymers are analyzed. Owing to the importance of the basic knowledge of the solution properties of polymers, the description of the conformational and thermodynamic behavior of polymers is presented in a classical way. The basic concepts like theta condition, excluded volume, good and poor solvents, critical phenomena, concentration regime, cosolvent effect of polymers in binary solvents, preferential adsorption are analyzed in an intelligible way. The thermodynamic theory of association equilibria which is capable to describe quantitatively the preferential adsorption of polymers by polar binary solvents is also analyzed.

Keywords Solution properties · Conformational analysis · Theta condition · Excluded volume · Good and poor solvent · Thermodynamic theories · Preferential adsorption · Cosolvent effect

1.1 Introduction: Solution Properties

Polymer solutions represent the most convinient systems for studying the properties of the macromolecules. In effect, almost the all information that we have now about the properties of macromolecules comes from the characterization realized in solution. This is the state in which linear chains are characterized. Osmotic pressure measurements in polymer solutions revealed for the first time the existence of high molecular masses and this result confirmed the macromolecular hypothesis. The development of our knowledge of the polymer solutions reflects to some extention the development of the Polymer Chemistry itself.

In a limited sense solutions are homogeneous liquid phases consisting of more than one substance in variable ratios, when for convenience one of the substances, which is called the solvent and may itself be a mixture, is treated differently from

L. Gargallo, D. Radić, *Physicochemical Behavior and Supramolecular Organization of Polymers*, DOI 10.1007/978-1-4020-9372-2_1,
© Springer Science+Business Media B.V. 2009

1

the other substances, which are called solutes [1]. Normally, the component which is in excess is called the solvent and the minor component(s) is the solute. When the sum of the mole fractions of the solutes is small compared to unity, the solution is called a dilute solution. A solution of solute substances in a solvent is treated as an ideal dilute solution when the solute activity coefficients γ are close to unity ($\gamma = 1$) [1, 2].

The deviations from ideal solution behavior are generally associated with a finite heat of solution. However, the properties of systems containing high molecular weight components, have shown extremely large deviations from the behavior to be expected of ideal solutions, even in cases where the heat of mixing was negligible.

To understand the thermodynamic behavior of a binary system containing a polymeric component and a low molecular weight component, it is necessary to consider that the most polymer molecules may be represented as flexible chains. If such chains are sufficiently long, the shape or conformation of their backbones may be likened to the random flight path of a particle undergoing Brownian motion and is then commonly refered to as a "random coil". The problem now is to analyze what happens with the shapes or conformations under different situations. At extreme dilutions, each one of these chains can assume a large number of conformations. The probability that any one chain exists at a given time in a given conformation will be independent of the conformations assumed by all the other chains. In the pure amorphous polymer the chain molecules are just as flexible as in solution. At the same time, it is possible to assume that they can be able to exist in a similar number of conformations. But, now these molecular conformations are not independent of each other. The shape of each molecular chain must be correlated with the shape assumed by its neighbors so as to fill the available space. When a molecular chain is transferred from the pure polymer phase to a dilute solution, this restraint is eliminated, and this accounts for the characteristic positive of the entropy mixing ΔS_M^E values of solutions of chain molecules. We can distinguish two ranges of concentration in systems containing chain molecules. In dilute solution, the polymer coils will only occasionally interpenetrate. At higher concentrations the total available volume is much less than the sum of the volumes enclosed by the twisting chain molecules. Then, in this range, the shape of a given chain, due to the presence of other polymer chains, will depend on the fraction of the volume occupied by these chains.

A quantitative theory of the change in conformational entropy produced by the mixing of flexible chain polymers with a solvent of low molecular weight was formulated by Flory [3] and Huggins [4].

In dilute solutions, the polymer chains behave, to a first approximation, as a gas. Indeed, the expression for the osmotic pressure is similar to the ideal gas law.

The "osmotic pressure of a solute" is the hydrostatic pressure that must be applied to a solution in order to increase the activity, a. (or fugacity, designated f, introduced by G. N. Lewis as a measure of thermodynamic "escaping tendency". It is an effective gas pressure corrected for deviations from the perfect gas laws) of the solvent sufficiently to balance its decrease caused by the presence of the solute. Equilibrium is established through a membrane permeable only to the solvent. This pressure is, by integrating

$$(dG/d\pi)_T = V \tag{1.1}$$

under the assumption of constant v_1 (negligible compressibility) and combining with (1.2)

$$G_1 - G_1^0 = RT \ \ln(f_1/f_1{}^0) \tag{1.2}$$

for the free energy of transfer of a mole of, for example, component 1 from pure liquid to solution. Whenever gas pressure obeys the ideal gas law with what is considered desired accuracy, fugacity can be replaced by gas pressure:

$$\pi = -RT/V_1 \ln f_1/f_1{}^0 = -RT/V_1 \ln a_1 \tag{1.3}$$

The osmotic pressure is a convenient variable for experiments, especially for high – polymer solutions.

There are, however, several intriguing facts that have aroused theoretical interest.

(i) the chain swells in good solvents, but does not in poor solvents (in the vecinity of a "Boyle" temperature.)
(ii) the chains overlap the total solution volume, while the polymer concentration is still low.

Thermodynamic predictions based on the liquid lattice theory do not fit osmotic experimental data [5].

For binary polymer-solvent, the Gibbs mixing function, ΔG_M, can be written, without approximation, as the sum of a combinatorial term plus an interactional term

$$\Delta G_M/RT = n_1 \ln v_1 + n_2 \ln \ v_2 + n_1 \ v_2 \ g_v \tag{1.4}$$

Here, n_i is amount of substance and v_i the volume fraction, this last magnitude being defined by $v_i = w_i v_{sp,i}/(w_1 v_{sp,1} + w_2 v_{sp,2})$, where w_i is the weight fraction and $v_{sp,i}$ the specific volume ($i = 1, 2$). Index 1 refers to solvent and index 2 to polymer. g is a phenomenological interaction parameter that takes into account deviations of ΔG_M from its combinatorial value. Subscript v in g_v denotes that g is defined on a volume fraction basis.

Differentiating equation (1.4) gives the chemical potentials of the components: $\Delta\mu_1$ and $\Delta\mu_2$. For the solvent

$$\Delta\mu_1/RT = \ln v_1 + (1 - V_1/V_2)v_2 + v_2^2\chi_v \tag{1.5}$$

Where

$$\chi_v = g_v + v_1(dg_v/dv_1) \tag{1.6}$$

V_i being molar volume and χ a phenomenological interaction parameter taking into account the deviations of $\Delta\mu_1$ from its purely combinatorial value. Subscript v in

χ_v denotes that χ is also defined on a volume fraction basis, the same as g_v. The equation (1.6) would be not strictly applicable to the dilute solution limit, but it can be interpreted as the definition of χ for the whole range of concentrations [6].

If instead of volume fractions, segment fractions, Φ_i, are used, then

$$\Delta G_M/RT = n_1 \ln \Phi_1 + n_2 \ln \Phi_2 + n_1 \Phi_2 g_\Phi \tag{1.7}$$

With $\Phi_i = w_i v_{sp,i}^* / \left(w_1 v_{sp,1}^* + w_2 v_{sp,2}^* \right)$, where $v_{sp,i}^*$ is the characteristic (hard – core) specific volume ($i = 1, 2$).

Differentiating equation (1.7) gives

$$\Delta u_1/RT = \ln \Phi_1 + (1 - V_1^*/V_2^*)\Phi_2 + \chi_\Phi \Phi_2^2 \tag{1.8}$$

Where

$$\chi_\Phi = g_\Phi + \Phi_1(dg_\Phi/d\Phi_1) \tag{1.9}$$

Subscript Φ on interaction parameters g_Φ and χ_Φ means that g_Φ and χ_Φ are defined on a segment fraction basis, and v_i^* is the characteristic molar volume. These $v_i^{*,5}$ are obtained from the reduced volumes, V_i

$$V_i = V_i/v_i^* \tag{1.10}$$

To obtain the reduced volumes, it is usual to use the equation of state due to Flory [7], from which is derived [7]

$$Vi = [1 + \alpha_i T/3(1 + \alpha_i T)]^3 \tag{1.11}$$

α_i being the thermal expansion coefficient.

For the polymer component, differentiation of equation (1.4) gives a result similar to equation (1.5)

$$\Delta \mu_2/RT = \ln v_2 + (1 - V_2/V_1)v_1 + (V_2/V_1)v_1^2 \chi_v'$$

Where

$$\chi_v' = g_v + v_2(dg_v/dv_2) \tag{1.12}$$

χ_v' being a phenomenological interaction parameter for the noncombinatorial part of the solute (polymer) chemical potential, defined on a volume fraction basis. Equations similar to equation (1.9) and (1.12) serve to define χ' on a segment fraction basis, χ_Φ'.

The relationship between the g parameter and the χ or χ' parameters is given by equations (1.4), (1.7), and (1.11). Integration of these equations up to the concentration $v_2(= 1 - v_1)$ or $\Phi_2(= 1 - \Phi_1)$ yields the value of g: g_v as function of v_2 or

g_Φ as a function of Φ_2. With the common symbol x to represent either ν or Φ, the results are

$$g_x = 1/x_1 \int_0^{x_1} \chi dx_1 = 1/x_2 \int_0^{x_2} \chi' dx_2 \qquad (1.13)$$

(x = ν or Φ). In the limit of zero concentration of polymer ($\nu_2 = \Phi_2 = 0$) and in the limit of pure polymer ($\nu_2 = \Phi_2 = 1$) we have

$$g_x^0 = \chi x^0 = \int_0^1 \chi dx_1 \qquad (1.14)$$

$$g_x^1 = \chi x^1 = \int_0^1 \chi' dx_2 \qquad (1.15)$$

where the superscripts 0 and 1 mean respectively $\nu_2 = \Phi_2 = 0$ and $\nu_2 = \Phi_2 = 1$.

Equations (1.13) and (1.14) show that the g interaction parameter is the reduced residual chemical potential (a) of the polymer, in the limit $\Phi_2 = 0$, and (b) of the solvent, in the limit $\Phi_2 = 1$ [6].

Theoretical g^0: The theoretical expression for the g^0 parameter, using the theory of polymer solutions developed by Flory and by Patterson based on the ideas of Prigogine and his school, has been given by Horta [8].

To calculate g^0, Masegosa et al. [6] have taken from the literature data of χ as a function of concentration. They have calculated g^0 for 41 polymer – solvent systems. The values of g^0 calculated are collected in Table 1.1.

In those cases in which $\tilde{V}_2/\tilde{V}_1$ is known, both g_ν^0 and g_Φ^0 are given. For the rest of the systems, only g_ν^0 is given. Prediction of thermodynamic properties on ternary systems formed by a polymer and two solvents or two polymers and a solvent requires the knowledge of the parameter g^0, characteristic of the interaction of the corresponding binary pairs [9]. However, due to the variety of sources for the several systems studied, the data correspond to different polymer molecular weights, m, and to different temperatures. Since the variation of χ with concentration may depend on M for low M's, it has selected data only for $M > 2 \times 10^9$, where no M dependence is detected.

Using the concept of a regular solution, it is possible to treat the free energy of mixing as being made up additively from contributions due to configurational probability and a free energy arising from nearest – neighbor interactions. The latter are characterized by the "Flory – Huggins interaction parameter", χ, which specifies, in units of RT, the excess free energy for the transfer of a mole of solvent molecules from the pure solvent to the pure polymer phase. With the initial state involving a solvent and a disordered polymer phase, the Flory – Huggins treatment leads to

$$\Delta G_M = RT(n_1 \ln \Phi_1 + n_2 \ln \Phi_2 + n_1 \chi \Phi_2) \qquad (1.16)$$

Table 1.1 Empirical values of the g^o interaction parameters at infinite dilution calculated from the experimental data of χ vs. polymer concentration. (From ref. [6])

System[a]	T, °C	$\bar{V}_2/\bar{V}_1$	$g_\phi{}^o$	$g_v{}^o$
PDMS–benzene	20, 25	0.9509[b]	0.65	0.63
PDMS–toluene[†]	20	0.9722	0.61	0.60
PDMS–cyclohexane[†]	20, 25	0.9517[b]	0.51	0.49
PDMS–n-pentane	20	0.9099	0.47	0.42
PDMS–n-hexane	20	0.9324	0.42	0.38
PDMS–n-heptane	20	0.9509	0.46	0.43
PDMS–n-octane[†]	20	0.9619	0.49	0.47
PDMS–n-nonane	20	0.9712	0.45	0.43
PDMS–2-2-4-trimethyl-pentane	20	0.9595	0.44	0.42
PDMS–3-methylpentane	20	0.9996	0.48	0.48
PDMS–p-xylene	20	0.9823	0.55	0.54
PDMS–ethylbenzene	20	0.9828	0.58	0.57
PDMS–hexamethyl-disiloxane	20	0.9303	0.34	0.29
PDMS–octamethyl-trisiloxane	20	0.9487	0.26	0.22
PS–cyclohexane[†]	20–30	0.8932[b]	0.84	0.82
	25		0.74	0.71[c]
PS–methyl ethyl ketone[†]	10, 25, 50	0.8817[b]	0.70	0.65
PS–ethylbenzene	10, 35	0.9211[b]	0.56	0.53
PS–diethyl ketone	20	0.8995	0.78	0.75
PS–acetone	25	0.8705	0.82	0.79
PS–n-propyl acetate	25	0.8813	0.71	0.67
PS–n-butyl acetate	20	0.9036	0.71	0.68
PS–benzene	15–45	0.8719[d]	0.42	0.34
			0.46	0.38
PS–toluene	25, 30	0.9221[b]	0.35	0.29
			0.42	0.37
PS–n-propyl ether	20	0.8904	0.82	0.80
PS–carbon tetrachloride	20	0.8891	0.45	0.38
PS–dioxane	20	0.9131	0.56	0.52
PIB–benzene[†]	25	0.8894	0.73	0.70
PIB–n-pentane[†]	25	0.8443	0.66	0.60
PIB–n-octane	25	0.8980	0.54	0.49
PIB–cyclohexane[†]	25	0.8901	0.48	0.42
NR–benzene[†]	25	0.9075	0.46	0.40
NR–methyl ethyl ketone	25	0.8965	0.83	0.81
NR–ethyl acetate	25	0.8924	0.84	0.82
PPO–carbon tetrachloride[†]	5.6	0.9391	−0.05	−0.12
PPO–chloroform[†]	5.6	0.9272	−0.86	−1.01
POCS–benzene	25, 40			0.55
POCS–methyl ethyl ketone	25			0.73
PP–diethyl ketone	25		0.85	
PP–diisobutyl ketone	25		0.70	
PBD–chloroform[†]	25		0.15	

[a] Dagger indicates data available on the whole concentration range. [b] At 25°C. [c] Reference [26]. [d] At 30°C.

The concept of "regular solution" was introduced by Hildebrand (1929) [10] and defined as a solution in which the partial molar entropies of the components are those to be expected from the ideal solution law. From this definition it follows that any deviation from ideal solution behavior in a regular solution is entirely accounted for by the heat of mixing. When Hildebrand first formulated the concept of regular solutions, he assumed that athermal solutions would necessarily follow the ideal solution law. Much later, when the physical chemistry of solutions of high molecular weight substances was subjected to detailed investigation, it became obvious that differences in molecular size of solute and solvent may lead to a very large deviation from solution ideality even if no heat effect accompanies the formation of the solution.

The entropy of mixing disoriented polymer and solvent may be obtained:

$$\Delta S_M{}^* = -k(n_1 \ln \Phi_1 + n_2 \ln \Phi_2) \tag{1.17}$$

An asterisk is appended to the symbol ΔS_M^* as a reminder that it represents only the configurational entropy computed by considering the external arrangement of the molecules and their segments. Contributions to the entropy resulting from specific interactions between neighbords will be considered later.

If the configurational entropy $\Delta S_M{}^*$ is assumed to represent the total entropy change ΔS_M on mixing, the free energy of mixing is obtained by combining equations

$$\Delta G_M = \Delta H_M - T\Delta S_M = \Delta H_M - T\Delta S_M{}^* \tag{1.18}$$
$$= kT[n_1 \ln \Phi_1 + n_2 \ln \Phi_2 + \chi n_1 \Phi_2]$$

The chemical potential μ_1 of the solvent in the solution relative to its chemical potential μ_1^0 in the pure liquid is obtained by differentiating the free energy of mixing, ΔG_M, with respect to the number n_1 of solvent molecules. Differentiation of equation for ΔG_M with respect to n_1 and multiplication of the result by Avogadro's number N in order to obtain the chemical potential per mole gives

$$\mu_1 - \mu_1^0 = RT[\ln(1 - \Phi_2) + (1 - 1/r)\Phi_2 + \chi \Phi_2^2] \tag{1.19}$$
$$r = V_2/V_1$$

This equation may be written

$$\mu_1 - \mu_1^0 = -TdS_1^* + RT \chi \Phi_2^2 \tag{1.20}$$
$$\mu_1 - \mu_1^0 = -TdS_1^* + \Delta H_1 \tag{1.21}$$

Where

$$dS_1^* = -R[\ln(1 - \Phi_2) + (1 - 1/r)\Phi_2] \tag{1.22}$$

is the relative partial molar configurational entropy of the solvent in the solution. It may be obtained directly by differentiation of equation (1.17). If x varies inversely with T, the first two terms in equation (1.22) represent the relative partial molar entropy.

$$\Delta S_M^i = -R(n_1 \ln x_1 + n_2 \ln x_2) \tag{1.23}$$

And the ideal partial molar entropy is obtained by differentiation with respect to n_1 or n_2

[with $x_1 = n_1/(n_1 + n_2)$ and $x_2 = n_2/(n_1 + n_2)$] as
$$\Delta S_1^i = -R \ln x_1; \ \Delta S_2^i = -R \ln x_2 \tag{1.24}$$

And if the solution is athermal, so that $\Delta H_1 = \Delta H_2 = 0$, the ideal free energy of mixing is

$$\Delta G_M^i = -T \Delta \Delta_M^i = RT(n_1 \ln x_1 + n_2 \ln x_2) \tag{1.25}$$

The solvent activity would then be given by

$$\ln a_1 = -\Phi_2 V_1/V_2 - 1/2(\Phi_2 V_1/V_2)^2 - 1/3(\Phi_2 V_1/V_2)^3 - \cdots + \chi\Phi_2^2 \tag{1.26}$$

In this case a large value of V_2/V_1 would make mixing impossible if χ had an appreciable positive value. Thus, endothermic mixing of high molecular weight polymers with solvents is possible only because of the conformational entropy gained by flexible chain molecules in the process of dilution.

Whatever the detailed interpretation of the thermodynamic behavior of polymer solutions, the term

$$(1/2 - \chi)\Phi_2^2 \tag{1.27}$$

in equation (1.28) arises from contributions to $\Delta G_1^E/RT$ due to binary interactions of the chain segments of the solute. These may have their origins in a change in the conformational entropy of the polymer, in changes in intermolecular contact energy in the mixing process, in a change in the randomness of orientation of solvent molecules when they are displaced from contact with the macromolecular solute, in volume changes...etc. Expressing by asterisks quantities resulting from such binary interactions, gives the relation:

$$1/2 - \chi = -(\Delta H_1^* - T\Delta S_1^*)/RT \ \Phi_2^2 \tag{1.28}$$

If we denote by Θ a temperature at which the coefficient of Φ_2^2 vanishes, then $\Delta H_1^* = \Theta \Delta S_1^*$, and this equation may be rewritten as

$$1/2 - \chi = \Psi(1 - \Theta/T) \tag{1.29}$$

where Ψ is

$$\Psi = T\Delta S_1^*/R\Phi_2^2 \qquad (1.30)$$

The use of the parameters Θ and Ψ has supplanted the interaction parameter χ, as suggested by Flory [11] which describes the behavior of a given polymer - solvent at a single temperature.

The assumption of forces of interaction between solvent and solute led to the century old principle that "like dissolves like". In many cases the presence of similar functional groups in the molecules suffices. This rule of thumb has only limited validity since there are many examples of solutions of chemically dissimilar compounds. For example, for small molecules methanol and benzene, water and N,N-dimethylformamide, aniline and diethyl ether, and for macromolecules, polystyrene and chloroform, are completely miscible at room temperature. On the other hand, insolubility can occur in spite of similarity of the two partners. Thus, polyvinylalcohol does not dissolve in ethanol, acetyl cellulose is insoluble in ethyl acetate, and polyacrylonitrile in acrylonitrile [12]. Between these two extremes there is a whole range of possibilities where the two materials dissolve each other to a limited extent.

Rather than the "like dissolves like" rule, it is the intermolecular interaction, between solvent and solute molecules, which determines the mutual solubility. A compound A dissolves in a solvent B only when the intermolecular forces of attraction K_{AA} and K_{BB} for the pure compounds can be overcome by the forces K_{AB} in solution [13].

The solubility parameter δ of Hildebrand [14] as defined in equation (1.31), can often be used in estimating the solubility of non-electrolytes solutes in organic solvents.

$$\delta = (\Delta E_v/Vm)^{1/2} = (\Delta H_v - RT/Vm)^{1/2} \qquad (1.31)$$

In this equation Vm is the molar volume of the solvent, and ΔE_v and ΔH_v are the molar energy and the molar enthalpy (heat) of vaporization for a gas of zero pressure, respectively. δ is a solvent property which measures the work necessary to separate the solvent molecules (i.e. disruption and reorganization of solvent/solvent interactions) to create a suitably sized cavity, large enough to accommodate the solute. Accordingly, highly ordered self-associated solvents exhibit relatively large δ-values. As a rule, it has been found that a good solvent for a certain non-electrolyte has a δ –value close to that of the solute [15].

When a polymer in solid state is in contact with a liquid solvent, we observe first a swelling phenomenon because the penetration of the small molecules of the solvent inside of the polymer structure. This behavior is different to that of the solutes non-macromolecules where the molecular identities are separated progressively to pass to the bulk of the solvent. In the case of the polymers this process is more complicated.

The mutual solubilities of components whose molecular sizes are drastically different is the case of the binary polymer-solvent systems, the molecules of the solute (polymer) are many order of magnitude larger than those of the solvent (monomer). In the solubility of the components whose molecular sizes are not significative different, the molar volume ratio is perhaps 2 or even 5, but always less than 10 [10].

The thermodynamic properties of polymers solutions have been reviewed by several authors [11, 17–19], we confine our attention here to the most common and perhaps also the most useful relation proposed by Flory [3] and Huggins [4] a generation ago.

The quantitative theory of the change in conformational entropy produced by the mixing of flexible chain polymers with a solvent of low molecular weight was formulated by Flory [3] and Huggins [4] who evaluated the number of distinguishable ways in which N_1 solvent molecules with a molar volume V_1 and N_2 polymers chains with a molar volume V_2 can be placed on a lattice so they each lattice site is occupy by either a solvent molecule or one of the V_2/V_1 segments of a polymer chain. In the calculation there is an assumption that, in placing a given chain segment on the lattice, which already contains previously placed chains, the probability of occupancy of a lattice site may be approximated by the overall fraction of occupied sites. This approximation is not real in very dilute solutions, where molecular coils, with a high local concentrations of chain segments, are separated by regions of pure solvent. The assumption of the Flory – Huggins theory is reasonable in the concentration range in which the chains interpenetrate each other, so that the density of chain segments is uniform, on the molecular scale, and it is in this range that the theory has been successful.

The Free energy change, ΔG, which results when we mix n_2 moles of polymer with n_1 moles of solvent at constant temperature and pressure is given by

$$\Delta G/RT = n_1 \ln \Phi_1 + n_2 \ln \Phi_2 + \chi \Phi_1 \Phi_2 (n_1 + V_2/V_1 n_2) \qquad (1.32)$$

Where V_2/V_1 (r) is the ratio of the molar volume of the polymer to that of the solvent and χ is the Flory parameter which depends primarily on the intermolecular forces between solute and solvent. According to the original formulation, this parameter is zero for athermal mixtures. However, subsequent work has shown that both the excess entropy and the excess enthalpy contribute to χ:

$$\chi = \chi_s + \chi_h \qquad (1.33)$$

where χ_s is the contribution from the excess entropy and χ_h is that from the excess enthalpy.

Knowledge of the magnitude of polymer – solvent interactions, and particularly of the "goodness" of a solvent for a given polymer, is very important for the investigation of the properties of polymers and the solutions and also for technological applications [20]. The goodness of solvents has hitherto been determined by either the Hildebrand solubility parameter δ [10] or the Flory – Huggins interaction parameter χ [21–25].

In the first case it is necessary to know solubility parameters of both the solvent, δ_1, and the polymer, δ_2. A general rule for non – polar systems is that the solvent is better when its δ_1 values is closer to δ_2. In polar systems, contributions of dispersive forces, dipole moments and hydrogen bonds to the total χ value [26–29] should be taken into account. The solubility parameter can thus be used only for a rough estimation of the goodness of a solvent, without claiming particular reliability of the conclusions drawn.

Another possible variable for the characterization of the goodness of solvents is the interaction parameter χ values, expressing the measure of deviations of actual solutions from ideal ones. This value can be determined by several methods, which are, mostly experimentally demanding and time-consuming. χ is dependent on both the polymer concentration and molecular weight and information provided about the specific interactions in the solution is of no particular interest [11, 30, 31]. Solvents, obviously different in quality, yield quite close values and thus the resolving capability is low. Comparison of results obtained by various methods and/or experimenters is thus fairly difficult [30, 32–34].

Once the second virial coefficient has been obtained for a given polymer – solvent system one can calculate the corresponding Flory – Huggins interaction parameter, χ, from the equation:

$$A_2 = (1/2 - \chi)/\rho_2^2 V_1 \tag{1.34}$$

Where ρ_2 is the density of the polymer (g cm^{-3}) and V_1 is molar volume of the solvent (cm^3 mol^{-1}).

We can also summarized a method for calculating the Flory – Huggins interaction parameter, χ, for a given polymer and solvent using the solubility parameters δ.

The solubility parameter of the polymer, δ_2, can be related to χ by: [35, 36]

$$\chi = V_1/RT(\delta_1 - \delta_2)^2 \tag{1.35}$$

where δ_1 is the solubility parameter of the solvent. (Units of solubility parameter are (energy/volume)$^{1/2}$, generally cal$^{1/2}$ cm$^{-3/2}$). The last equation can be rewritten as

$$\delta_1^2/RT - \chi/V_1 = [2\delta_2/RT]\delta_1 = \delta_2^2/RT \tag{1.36}$$

This is the equation of a straight line. Hence, when the left – hand side is plotted as a function of δ_1 one can estimate δ_2.

Figure 1.1 is an example of a plot obtained from equation (1.34). The δ_2 value estimated is 10.2 cal$^{1/2}$ cm$^{-3/2}$ [37]. The values of R and T used were 1.99 cal/mol K and 298 K, respectively.

There are also other quantities that are dependent on the goodness of solvents. Among them is the Huggins viscosity constant k', which can be determined quite easily and, because of its interesting properties, seems to be suitable for direct determination of the goodness of a particular solvent [20].

The dependence of viscosity η of dilute polymer solutions on concentration c can be described by a polynomial in the form [31, 38].

$$\eta = \eta_0(1 + a_1 c + a_2 c^2 + \ldots) \tag{1.37}$$

where η_0 is the viscosity of the pure solvent. This equation is generally presented in the form:

Fig. 1.1 Plot of
equation (1.36) for poly(vinyl
acetate) in benzene (1),
Chloroform (2),
Chlorobenzene (3), methyl
ethyl ketone (4), acetone (5)
and acetonitrile (6). (From
ref. [37])

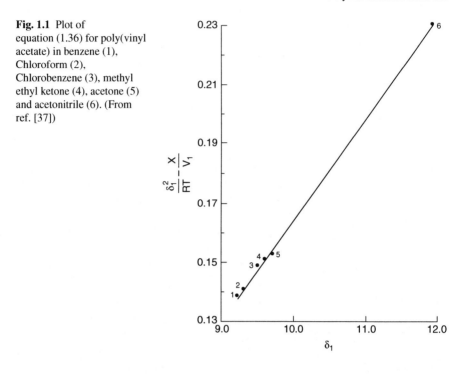

$$[\eta] = \eta_0(1 + [\eta]c + k'[\eta]^2c^2 + \ldots) \tag{1.38}$$

where $[\eta]$ is the intrinsic viscosity and k' is the dimensionless Huggins viscosity constant. Neglecting terms with third and higher powers of concentration yields the well known Huggins equation. Figure 1.2 shows a classical plot to obtain $[\eta]$ and k' for several fraction of the poly (monobenzyl itaconate) (PMBzI) [39].

As mentioned in some monographs [31, 38] and confirmed by numerous experimental results [40–43], the Huggins constant is independent of the molecular weight of the polymer. Its value is influenced only by the goodness of the solvent. However, it can be expected that the Huggins constant will be molecular – weight – dependent only in polymers easily associating in solution, either by the effect of specific interactions as strong ionic, or polar interactions or by the effect of hydrogen bonds [20].

One of the most surprising generalities in the world of polymers is that $[\eta]$ values for a series of homologous polymers under a fixed solvent condition (solvent and temperature) follows a simple power law as

$$[\eta] = KM^\nu \tag{1.39}$$

over an extended range of M. Here, K and ν are constants for the polymer + solvent considered. This equation (1.39) is referred to as the Mark – Houwink – Sakurada (MHS).

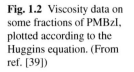

Fig. 1.2 Viscosity data on some fractions of PMBzI, plotted according to the Huggins equation. (From ref. [39])

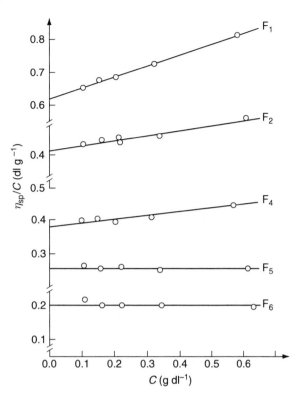

The relation between number molecular weight, Mn and intrinsic viscosity, $[\eta]$, for poly(pentachlorophenyl methacrylate) (PPClPh) can be represented by the Mark – Houwink – Sakurada equation [44].

Figures 1.3 and 1.4 ilustrate these double logarithmic plots in different solvents. Becerra et al. [44] have found for PPClPh, the following relations:

o-Dichlorobenzene at 25°C:	$[\eta] = 25{,}4 \cdot 10 - 5\,Mn^{0{,}67}$
o-Xilene at 25°C:	$[\eta] = 28{,}6 \cdot 10 - 5\,Mn^{0{,}63}$
Chlorobenzene at 25°C:	$[\eta] = 29{,}1 \cdot 10 - 5\,Mn^{0{,}63}$
Toluene at 25°C:	$[\eta] = 35{,}2 \cdot 10 - 5\,Mn^{0{,}58}$
Benzene at 40°C:	$[\eta] = 53{,}7 \cdot 10 - 5\,Mn^{0{,}50}$
Ethylbenzene at 25°C:	$[\eta] = 61{,}0 \cdot 10 - 5\,Mn^{0{,}50}$

The results obtained on poly(pentachlorophenyl methacrylate) show that $[\eta]$ is accurately proportional to $Mn^{0{,}50}$ for the ideal or theta (θ) solvent.

According to Fugita [45] the main experimental facts that have to be explained theoretically are as follows:

1. When $[\eta]$ is plotted against M on a log – log graph paper, it gives a straight line over a wide range of M;

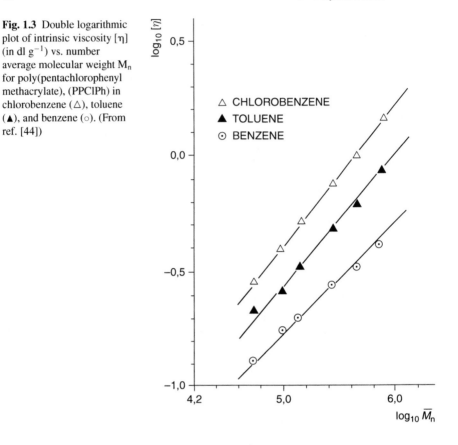

Fig. 1.3 Double logarithmic plot of intrinsic viscosity [η] (in dl g^{-1}) vs. number average molecular weight M$_n$ for poly(pentachlorophenyl methacrylate), (PPClPh) in chlorobenzene ($\triangle$), toluene ($\blacktriangle$), and benzene ($\odot$). (From ref. [44])

2. The slope ν of the line for linear flexible polymers in non $-\theta$ solvents in which the second virial coefficient A$_2$ is positive is in the range $0.5 < \nu < 0.8$.
3. In general ν is larger for a better solvent, i.e., for a larger A$_2$;
4. Under the θ condition where A$_2$ vanishes, ν for flexible linear polymers is always 0.50.

Several theoretical tentatives have been proposed to explain the empirical equations between [η] and M. The effects of hydrodynamic interactions between the elements of a Gaussian chain were taken into account by Kirkwood and Riseman [46] in their theory of intrinsic viscosity describing the permeability of the polymer coil. Later, it was found that the Kirdwood – Riseman treatment contained errors which led to overestimate of hydrodynamic radii R$_\eta$. Flory [47] has pointed out that most polymer chains with an appreciable molecular weight approximate the behavior of impermeable coils, and this leads to a great simplification in the interpretation of intrinsic viscosity. Substituting for the polymer coil a hydrodynamically equivalent sphere with a molar volume V$_e$, it was possible to obtain

$$[\eta] = 5/2V_e/M_2 = \Phi' < S^2 >^{3/2} /M_2 \qquad (1.40)$$

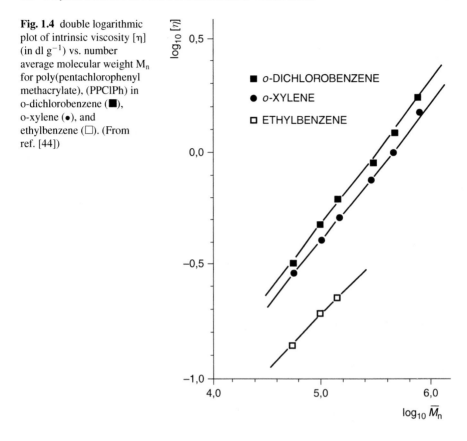

Fig. 1.4 double logarithmic plot of intrinsic viscosity [η] (in dl g^{-1}) vs. number average molecular weight M_n for poly(pentachlorophenyl methacrylate), (PPClPh) in o-dichlorobenzene (■), o-xylene (•), and ethylbenzene (□). (From ref. [44])

where $\Phi' = 10/3\pi N(R_\eta/<S^2>^{1/2})^3$ should be a universal constant independent of the nature of the macromolecule and independent of the solvent medium.

The relation (1–40) leads to a number of interesting consequences. In a theta solvent, in which the shape of the chain is described by the random flight model, $<S^2>$ is proportional to M_2, so that the intrinsic viscosity should be proportional to $M_2^{1/2}$. And this prediction has been applied and verified. In solvent media better than θ-solvents, the theory of Flory [11,46] predicts that the linear expansion factor a increases for any polymer – homologous series with chain length. Thus the exponent ν in the empirical equation should be larger than 0.50.

1.2 Polymer Solutions in Good Solvent: Excluded Volume Effect

If we suppose a very dilute gas of random flight chains. As a result of thermal rotation of chain segments each chain will take up a great number of conformations, with a very short interval of time being spent for the passage from one conformation to another. In so doing, it automatically avoids taking those conformations in which any pair of its segments (beads) overlap, because it is physically impossible for them to occupy the same volume element in space at the same time. When a pair of segments (beads) come close they exert a repulsive force F on each other. The strength of this force depends on the separation between the segments (beads) and also on the chemistry of each segment (bead), as well as the temperature and pressure of the gas. Now, if we introduce a liquid solvent into the system to make a dilute solution of random flight chains, each segment (bead) has a chance to interact with solvent molecules as well as other segments (beads). As a result the force that acts between a pair of segments (beads) becomes no longer equal to the vacuum value F. If the segment – solvent (bead – solvent) interaction favors segment – solvent contact over the segment - segment one, the solvent is said to be good (for the polymer considered), while if the reverse is the case, the solvent is said to be poor or bad. Thus, a good solvent tends to prevent a pair of segments (beads) from approaching or tries to pull them apart. This suggests that the segment – solvent (bead – solvent) interaction has the effect equivalent to inducing some additional force between a pair of segments (beads) [45].

According to the statistical mechanical theory formulated by McMillan and Mayer [48] and by Saito [49], the solution of a polymer in a pure solvent should be equivalent in thermodynamic behavior to a hypothetical gas of the same polymer whose segments repel each other with a force given by the sum of the vacuum value F and some additional force F'. The latter summarizes the effect due to segment – solvent interactions and is called the solvent – mediated force. By this way, in the McMillan – Saito theory, the segment – solvent interactions are lumped into an unknown F', and the thermodynamics of polymer solutions can be formulated by applying the well – established theory of gases. However, it is necessary to note that, stritly speaking, this remarkable advantage can be used at a fixed chemical potential of the solvent [45].

In a good solvent, the segment – solvent interaction tends to pull a pair of segments apart, so that the solvent – mediated force F' should be repulsive as is F. On the other hand, F' should be attractive in poor solvents. Hence, as the solvent is made poorer by changing either solvent species or temperature, the situation should be reached in which the attractive F' cancels or suppresses the repulsive F so that the net force F + F' becomes zero or even negative.

What it is concerned with in the physics of polymer systems is not their physical properties for individual polymer conformations but those averaged over the ensemble n of such conformations under given conditions of polymer and solvent. For flexible polymers the averages depend primarily on the strength of the net interaction force F + F' and the number of segments contained in the chain. Knowledge of F and F' is essential for the understanding of polymers in solution.

The above discussion is concerned with the interactions between a pair of segments, i.e., binary cluster interactions. However, there are opportunities for the segments (beads) to interact forming clusters higher than the binary one. Flory was the first to reach a very important recognition that polymer conformations and global polymer properties are significantly influenced by the potential energy stored in the polymer chain as a result of formation of binary, ternary, and higher – order clusters of segments.

The term **excluded – volume effect** is used or described any effect arising from intrachain or interchain segment – segment interactions. This interaction, which Flory referred to as a long – range interference of monomer units, decidedly affects the number of conformations that the chain can take up; for example, those in which two monomers occupy the same point in space are not realizable. Flory's recognition of this effect triggered the development of polymer solution studies for the last four or five decades.

1.3 Theta Condition: Concentration Regimes

The osmotic pressure π of a dilute solution of a monodisperse polymer with molecular weight M is expressed as a power series of polymer mass concentration c (weight of polymer per unit volume of solution) as

$$\pi/RT = c/M + A_2c^2 + A_3c^3 + \dots \tag{1.41}$$

Where R is the gas constant, T the absolute temperature, and M the molecular weight of the polymer. This series is usually called the osmotic virial expansion, with $A_i(i = 2, 3, \dots)$ being referred to as the i-th virial coefficient of the solution.

Experimentally, A_2 can be evaluated by determining the initial slope of $\pi/(RTc)$ plotted against c. However, the estimation of A_3 and the higher virial coefficients is not a simple matter for experimentalists. Available experimental information about the virial coefficients is largely limited to A_2.

For a series of homologous polymers A_2 depends on M as well as T and the nature of the solvent. Experimental studies have repeatedly shown that for a given polymer there is a combination of poor solvent and temperature Θ for which A_2 vanishes regardless of M. This spetial poor solvent at Θ is called the theta solvent, and Θ, the theta temperature.

Figure 1.5 shows the results reported for Poly(pentachlorophenyl methacrylate) in benzene solution at 40°C studied by osmotic pression [44]

The osmotic pressure function π in benzene at 40°C is independent on polymer concentration c. This result is a proof that this solvent is a Θ – solvent ($A_2 = 0$) for poly(pentachlorophenyl methacrylate). The results are shown in Figure 1.5 for three fractions.

Another osmometric data are given in Table 1.2 for this polymer in toluene at 25°C.

Osmometric measurements were also carried out for different fractions of poly [4-(1,1,3,3-tetramethylbutyl)phenyl methacrylate] in toluene at 25°C [50]. The measurements were made at five or six different concentrations and extrapolated to infinite dilution. Figure 1.6 shows the plot π/c versus c, which agrees with classical relations [51,52]

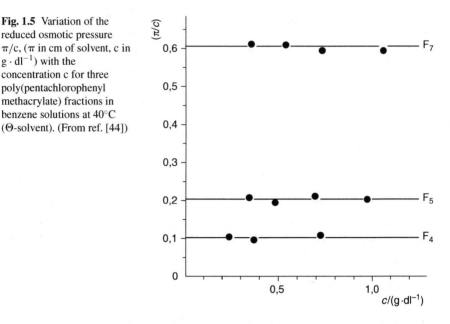

Fig. 1.5 Variation of the reduced osmotic pressure π/c, (π in cm of solvent, c in $g \cdot dl^{-1}$) with the concentration c for three poly(pentachlorophenyl methacrylate) fractions in benzene solutions at 40°C (Θ-solvent). (From ref. [44])

The polymer studied in the series of poly(methacrylic ester)s with a mesogenic side group was polymer labeled 1a shown in Scheme 1.1

Table 1.3 summarizes molecular weights (M_n) and second virial coefficient (A_2) data for various fractions of this polymer.

The variation of A_2 with the molecular weight shows a regular behavior [53]. Figure 1.7 shows this result [50].

Toluene is a good solvent for this polymer.

Sufficiently dilute polymer solutions may be viewed as systems in which "islands" of polymer coils scattered in the "sea" of a liquid solvent occasionally impinge and interpenetrate. By this way, the spatial distribution of chain segments in them is quite heterogeneous and undergoes appreciable fluctuations from time to time. As the polymer concentration increases, the collision of the islands becomes more frequent and causes the chains to overlap and entangle in a complex fashion.

Table 1.2 Reduced osmotic pressures $(\pi/c)_{c=0}$, number average molecular weights M_n and osmotic second virial coefficient A_2 for poly(pentachlorophenyl methacrylate) fractions in toluene at 25°C and benzene at 40°C (π in cm of benzene or toluene) (c in $g \cdot dl^{-1}$). (From ref. [44])

Fractions	$\left(\dfrac{\pi}{c}\right)_{c=0}$ [a]	$\bar{M}_n \cdot 10^{-5}$ [a]	$\left(\dfrac{\pi}{c}\right)_{c=0}$ [b]	$\bar{M}_n \cdot 10^{-5}$ [b]	$\dfrac{A_2 \cdot 10^{4}\,[b]}{cm^3 \cdot mol \cdot g^{-2}}$
F_1	0,45	6,83	0,40	7,21	2,26
F_2	0,70	4,40	0,67	4,34	2,38
F_3	–	–	0,92	3,18	3,34
F_4	1,06	2,90	1,02	2,85	3,19
F_5	2,11	1,45	2,12	1,38	2,30
F_6	3,32	0,93	3,03	0,96	3,91
F_7	6,05	0,51	5,28	0,55	2,40

[a] In benzene at 40°C.
[b] In toluene at 25°C.

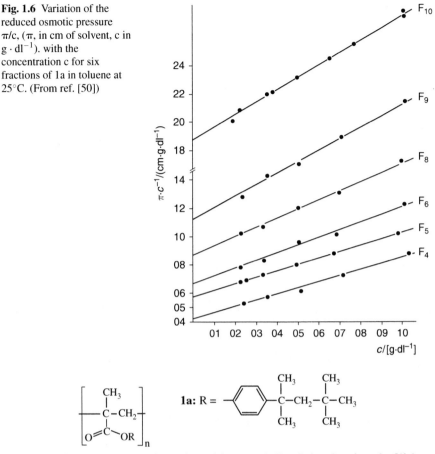

Fig. 1.6 Variation of the reduced osmotic pressure π/c, (π, in cm of solvent, c in $g \cdot dl^{-1}$). with the concentration c for six fractions of 1a in toluene at 25°C. (From ref. [50])

Scheme 1.1 Chemical structure of poly[4-(1,1,3,3-tetramethylbutyl)phenyl methacrylate](1a).

As a result the segment distribution becomes less heterogeneous and its fluctuation is by and large suppressed. Increased chances of chain contact may lead to interchain association.

Table 1.3 Reduced osmotic pressures $(\pi/c)_{c=0}$, number average molecular weights M_n and osmotic second virial coefficient A_2 for fractions of poly(1,1,3,3-tetramethylbutylphenyl methacrylate)(1a) in toluene at 25°C. (From ref. [50])

Fractions	$(\pi/c)_{c=0}$	$10^{-5} \cdot \overline{M}_n$	$\dfrac{10^4 \cdot A_2}{cm^3 \cdot mol \cdot g^{-2}}$
F_3	0,42	7,09	1,7
F_4	0,57	5,07	1,8
F_5	0,67	4,37	2,2
F_6	0,87	3,35	2,6
F_7	1,13	2,59	3,2
F_8	1,87	1,56	3,7

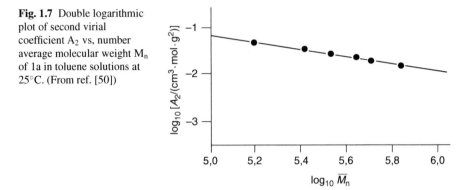

Fig. 1.7 Double logarithmic plot of second virial coefficient A_2 vs, number average molecular weight M_n of 1a in toluene solutions at 25°C. (From ref. [50])

One of the characteristic phenomena which occur when a polymer solution in a non – Θ solvent is concentrated is the "screening effect" on chain dimensions. This is a change in $<S^2>$ (c) toward $<S^2>_\Theta$ with the increase in polymer concentration. Here, $<S^2>$ (c) is the mean – square radius of giration of a single polymer chain at a mass concentration c, and $<S^2>_\Theta$ is its infinite – dilution value under the Θ condition. Daoud et al. [54] were the first to observe the screening effect. Studying polystyrene in carbon disulfide by small – angle neutron scattering (SANS), they found $<S^2>$ (c) to decrease in proportion to $c^{-0.25}$.

Figure 1.8 [45] shows a demonstration of the screening effect by King et al. [55] who investigated polystyrene in deuterated toluene by SANS; the abscissa b denotes the polymer concentration in terms of the mole fraction of styrene residues.

The slope of the indicated line is – 0.16, differing from – 0.25 obtained by Daoud et al. [54] The screening effect is a reflection of the decrease in intrachain segment – segment interactions caused by chain overlapping.

If we look at a single chain in a polymer solution, the average number density of its segments, η_{coil}, i.e., their number contained in unit volume of the average space occupied by the chain, is given by

$$\eta_{coil} = 3M/m/4\pi[< S^2 > (c)]^{3/2} \tag{1.42}$$

where M is the molecular weight of the polymer and m the molar mass of one segment. If we denote the average number density of chain segments in the entire solution by η_{soln}, we obtain

$$\eta_{soln} = (c/m)N \tag{1.43}$$

where N is the Avogadro constant. For dilute solutions in which the "islands" are separated by the solvent sea we have $\eta_{soln} < \eta_{coil}$ even though polymer coils occasionally overlap. With increasing c we reach the point at which η_{soln} catches up with η_{coil}. Denoting the c value for this solution by c*, we have the relation

$$c^* = 3M/4\pi N_A[< S^2 > (c^*)]^{3/2} \tag{1.44}$$

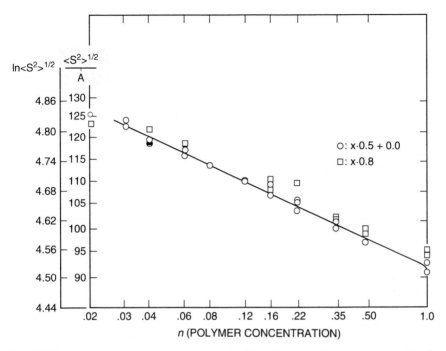

Fig. 1.8 Concentration dependence of coil radius for PS in d-toluene. Line, least-squares fit to the sample pairs for x = 0.50 and 0.0. Here, n is the mole fraction of styrene residues, and x the mole fraction of d-styrene in a mixture of h- and d- styrenes. (From ref. [45])

Usually, c^* is called the overlap concentration. It would correspond to the situation where polymer coils would begin to touch one another throughout the solution if they behaved like mutually impenetrable spheres. However, having an open structure, actual polymer coils start interpenetrating as soon as the solution leaves the state of infinite dilution. Hence, it is not legitimate to consider c^* as if marks the onset of coil overlapping according to Fugita [45]. The term overlap concentration thus seems misleading.

c^* is often approximated by c_0^* defined as

$$c_0^* = 3M/4\pi N_A[< S^2 > (0)]^{3/2} \qquad (1.45)$$

With $<S^2>(0) \sim M^{2\nu}$ valid unless M is low, this equation gives

$$c_0^* \sim M^{-(3\nu-1)} \qquad (1.46)$$

Which predicts

$$c_0^* \sim M^{-0.5}(\Theta \text{ solvents}) \qquad (1.47)$$

$$c_0^* \sim M^{-0.8}(\text{good solvents}) \qquad (1.48)$$

Really, there is no established consensus about the definition of the overlap concentration. In fact, Hager and Berry [56] replace the factor $3/4\pi$ by 1/5, and Graessley [57] by 1/8, while Ying and Chu [58] drop this factor and use $<R^2>(0)/4$ for $<S^2>(0)$. Thus we have to be careful in comparing reported overlap concentrations.

It is important to note that c^* is not a critical concentration. It should not expect that something special, such as sharp changes in the concentration dependence of some physical properties of the solution, takes place at this concentration. However, it seems certain that the macroscopic distribution of chain segments over the entire solution becomes essentially uniform when c passes through a relatively narrow region around c^*. Thus, it is possible to take c^* as a measure of the crossover region where the island – sea heterogeneous structure of a polymer solution changes to the state of macroscopically uniform segment distribution.

At concentrations higher than c^* the latter state remains unchanged, and the average segment density can be equated to c. However, when viewed microscopically, even the solutions above c^* are not uniform, the segment density in each volume element fluctuating from time to time about the mean value c and those in different volume elements at a given instant being different. Because of the chain connectivity of segments and their intra – and interchain interactions the density fluctuations at different places in the solution cannot take place independently. In other words, they are correlated. This correlation governs in several ways the physical properties of concentrated polymer solutions. Thus, this is a key concept [45] in the discussion of the concentrated regime.

Some authors [59] call the region $c < c^*$ the virial regime. Probably they consider that the virial expansion for osmotic pressure would diverge as c approaches c^*.

In dilute solutions, the chains are far apart on average. When the polymer concentration c increases, there exists a concentration c^* at which the chains begin to overlap. This is the onset of the semi-dilute regime. It may write:

$$c^* \propto 1/(R^2)^{3/2}$$

Where R^2 is the square end-to-end distance at zero concentration.

The concentration c^* can be very low if the chains are long. Indeed, in the limit of infinitely long chains, $c^* = 0$. Thus a given concentration interval Δc can either belong to the dilute or to the semi-dilute regime, depending on the length of the chain. The main difference between a dilute and a semi – dilute polymer concentrations is the homogeneity of the polymer distribution in space. Homogeneity is a discriminating factor for the effects of repulsive interaction. Dilute solutions have a heterogeneous structure: the polymer chains form isolated islands in the solvent. The repulsive interactions between monomers add up to swell the chains.

Solutions with overlap (semi-dilute) possess a homogeneous structure on the large scale ($\sim$R). However, in between nearest neighbour contacts, the solution is dilute, and therefore inhomogeneous. The repulsive interactions acting on the homogeneous structure do dent of the polymer size not combine to swell the structure. On the contrary, compensations occur and on average the interactions screen the pair correlation function. This means that correlations between monomers are effectively weaker than those associated with a random walk. Screening in polymer solutions was introduced by Edwards [60] and proved to be a unifying concept. In semi-dilute solutions, screening and swelling coexist, at different scales. This generates a structure. Chain sequences swell only within distances §, the screening length. This length depends on monomer concentration and interaction strength (it is independent of polymer size).

Because screening and swelling coexist, the length § varies in a singular manner with concentration. This is based on the following remarks.

(i) An intrinsic measure of the monomer concentration is the "Kuhnian" concentration [60]

$$c_k = c(R^2/3)^{1/2v}(\text{dimension } L^{1/v-d}) \tag{1.49}$$

This concentration defines the "Kuhnian" overlap length [60]

$$\S_k = c_k^{-1/(d-1/v)} \tag{1.50}$$

(ii) The screening length § is proportional to $\S_k$

$$\S = \Gamma\S_k \tag{1.51}$$

where Γ is a universal constant [61] approximated by the relation (1.52)

$$\Gamma = (1/4\pi)(g^*)^{-1/2} = 0.165 \tag{1.52}$$

derived from the simple tree approximation of the structure function.

Experimental values of § are obtained from the scattering intensity data I (q) and formula

$$C/I(q) = A(q^2 + \S^{-2}) \tag{1.53}$$

Where A is a constant. The singularity $q^2 = -\S^{-2}$ implies screening in real space. i.e. an attenuation factor $e^{-r/\S}$.

The semi-dilute regime exists in the limits of infinite chains, since the singularity is independent of chain size [61].

The values of § derived in this manner satisfy relation (1.51). The experimental value of Γ is found to be $\Gamma = 0.18 \pm 0.015$ [61].

1.4 Critical Phenomena in Polymer Solution: The Collapse of Macromolecules in Poor Solvents

Under appropriate conditions of temperature, concentration and solvent composition most substances aggregate and precipitate frequently, producing a crystalline phase. In the case of long – chain molecules, in particular synthetic polymers, the phase resulting from aggregation is usually a concentrated solution of disordered macromolecules. The coexisting phase is very dilute solution in which the chains are more contracted than in the concentrated phase, sometimes reaching the state of very compact globules [62]. This collapsed or globular state is very difficult to investigate by scattering methods, for the twofold reason that the scattering signal is very faint owing to the low concentration and that the kinetics of chain collapse may be so slow that equilibrium is difficult to establish. In spite of these difficulties, in a few cases macromolecular collapse was observed as a function of temperature; accurate measurements of the collapse kinetics of long polystyrene chains were reported.

Some decades ago Stockmayer [63] first suggested that a flexible polymer chain can transit its conformation from an expanded coil to a collapsed globule on the basis of Flory's mean – field theory [11]. Since his prediction, theoretical and experimental studies of this coil – to –globule transition have been extensively conducted [31, 64–68].

The majority of the reported works were based on polystyrene in various organic solvents because of two very special requirements for the studied polymer chain; namely, its molecular weight should be as high as possible and its molecular mass distribution should be as narrow as possible. Many useful experimental results have been obtained by using improved modern equipments [69, 70].

The great interest in this coil – to – globule transition is not only due to its importance as a general and fundamental concept in polymer physics and solution dynamics but also due to its deep implications in many biological systems, such as protein folding [71] and DNA packing [72].

If the solvent quality improves with increasing temperature macromolecular aggregation will occur at temperatures lower than a critical temperature $T_C < \Theta$ (the "$_c$" subscript stands for "critical point"). Figure 1.9 [62] shows one example.

The main features of the polymer – solvent phase diagram can be obtained at the simple Flory – Huggins level [11, 73] In effect, this theory leads to the following predictions for the dependence of the position of the critical point on the molecular mass $(M \to \infty)$:

$$\Phi_C \propto M^{-1/2} \tag{1.54}$$

$$[\Theta - T_C/\Theta] \propto M^{-1/2} \tag{1.55}$$

Φ being the polymer volume fraction. Thus, besides separating the regimes of chain collapse and expansion, the Θ – temperature has characterized by $\chi > 1/2$. According to equation (1.54), the limiting condition $\chi = 1/2$ corresponds to $T = \Theta$. This condition may be attained either by a variation in temperature for solvents

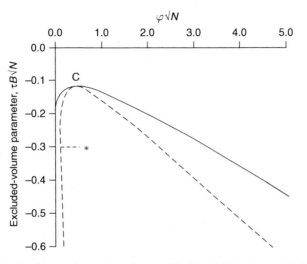

Fig. 1.9 Calculated polymer-solvent phase diagram. The bimodal *(continuous line)* is the coexistence curve; the points below it correspond to thermodynamically unstable states, which undergo phase separation. However, the pints between the bimodal and the spinodal *(dashed line)* are kinetically stable, since there is a free-energy barrier to phase separation. "C" indicates the critical point "*" the collapse temperature. The deviation of the low-concentration branch of the spinodal from the vertical axis below T* is an artifact of the mean-field approximation. (From ref. [62])

of constant composition or by a variation in the composition of a mixed solvent medium at constant temperature.

Combination of equations (1.54) and (1.55) leads to

$$1/T_c = 1/\Theta[1 + 1/\Psi(V_1/V_2)^{1/2} + V_1/2V_2] \qquad (1.56)$$

another important physical meaning: it is the upper limit for phase separation or, in other words, the upper critical solution temperature (UCST).

In the opposite case of solvent quality improving with decreasing temperature, which may occur with polar systems, the Θ – temperature is the lower critical solution temperature (LCST). The critical temperature is always bracketed by the Θ – temperature and the collapse temperature (T^* < T_C < Θ in the UCST case, vice versa in the LCST case).

The interplay and competition between chain collapse and phase separation leads to several challenging questions. Results on the solution properties in the neighbourhood of the critical point have been reviewed by Sanchez [74] and Widom [75] One mayor conclusion following from the experimental investigations is that the critical volume fraction has a weaker dependence on the chain length than the classical prediction. One possible implication of this result is that the critical point represents a new scaling regime for the chain statistics, quite distinct both from the Θ – state and from the fully collapsed globule.

Experimental investigations on the critical phenomena in a polymer solution have been made extensively with interests mainly in the universality of the critical exponents which indicates that the critical exponents do not depend on any details of materials as long as the critical phenomena belongs to the same symmetry class. On the other hand, the critical amplitudes are dependent on details of materials; then, in a polymer solution, little attention was paid to them except the correlation length. The molecular weight dependence of the correlation length has been investigated since Debye [76] proposed a theory of critical opalescence in a polymer solution. Experimental investigations were made by Debye et al. [77] and by Chu [78] and they found that

$$l^2 \propto M^{0.57} \tag{1.57}$$

where l is the Debye length [76] and M is the molecular weight, although the Debye theory predicts $l^2 \propto M$. Theoretical attempts were also made to explain the disagreement by de Gennes [79] and by Vrij and Esker [80]. They showed that l should be proportional to $M^{1/4}$ within the mean field treatment.

Allegra et al. [62] have carried out an explicit calculation of the phase diagram, starting from a simple mean – field expression for the solution free energy as a function of temperature, polymer volume fraction and radius of gyration of the chains [81]. The result is shown in Fig. 1.9, for a choice of the chain parameters roughly corresponding to atactic polystyrene: the continuous and the dashed lines represent the binodal and the spinodal [11, 73]. In this approach, the critical – point scaling laws are identical to the classical Flory – Huggins ones.

It is interesting to mention a theoretical studied reported by Fields et al. [82] on the collapse and aggregation of proteins and random copolymers. They concluded that the form of the phase diagram is extremely sensitive to the copolymer composition and that, compared with the homopolymer case, copolymer solutions are much more stable against aggregation in otherwise analogous conditions, precipitation occurs at concentrations that can be several orders of magnitude higher.

The phase separation behavior of a sample of poly(N-vinyl-2.pyrrolidone) (PVP) in aqueous Na_2SO_4 (0.55 M) containing a surfactant, sodium dodecyl sulfate (SDS), was reported [83]. This phenomenon was studied in function of temperature and surfactant concentration as independent variables. Without surfactant, the polymer exhibits a lower solution temperature (LCST) of 28°C, above which it precipitates. At SDS concentrations of only 300 mg L^{-1}, aggregation was prevented and the behavior of isolated polymer molecules could be studied.

Measurements of hydrodynamic radius (R_H) and intrinsic viscosity [η] of PVP in Na_2SO_4 (0.55 M) have been performed by using Dynamic Light Scattering (DLS) and viscometry. From these results, a coil – to – globule phase transition was detected.

Figure 1.10 shows the phase behavior of the PVP-SDS system.

Table 1.4 shows the data found for PVP fraction studied as a function of temperature and SDS concentration [83].

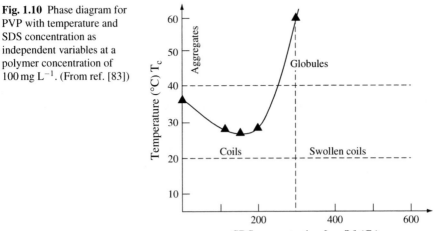

Fig. 1.10 Phase diagram for PVP with temperature and SDS concentration as independent variables at a polymer concentration of 100 mg L^{-1}. (From ref. [83])

Mutually compatible polymers are relatively rare [84, 85]. A typical phase diagram of two incompatible polymers and a common solvent is that determined by Kern for the polystyrene – poly(p – chlorostyrene) – benzene system shown in the Fig. 1.11 [86].

It may be seen that benzene, which is miscible in all proportions with either polymer alone, can dissolve only slightly more than 2% of a mixture of equal weights of the two polymers without phase separation. Poly(p–chlorostyrene) has a negligible solubility in a phase containing 4% of polystyrene. Kern used a polystyrene sample with a substantially longer chain length than that of poly(p–chlorostyrene), and this accounts for the asymmetry of the phase diagram. It is generally found that the mutual incompatibility of polymers increases rapidly with an increase in their molecular weight.

The thermodynamic causes of this phase separation phenomenon are not difficult to visualize [19]. If two dilute solutions of low molecular weight solutes A and B in the same solvent medium are mixed, the system forms a single phase, since the gain in entropy will outweigh even an unfavorable energy of mixing. However, if

Table 1.4 Intrinsic viscosity [η], hydrodynamic radius R$_H$ and linear expansion coefficient α_n for PVP in aqueous Na$_2$SO$_4$ (0.55 M) at different temperatures and SDS concentration. (From ref. [83])

State	[η] (dL g^{-1})	R_h (from DLS, nm)	α_η [a]
Aggregate	–	~900.0	–
Single-globule	0.17[b]	28.2	0.74
Coil	0.41[c]	88.9	1.00
Swollen coil	1.06[d]	120	1.37

[a] α_η was calculated taken as [η]$_\theta$ = 0.41 dL g^{-1} at 20°C.
[b] At 40°C and C_{SDS} 300 mg L^{-1}.
[c] At 20°C and C_{SDS} 300 mg L^{-1}.
[d] At 20°C and C_{SDS} 600 mg L^{-1}.

Fig. 1.11 Phase diagram of
two incompatible polymers
and a common solvent. (From
ref. [86])

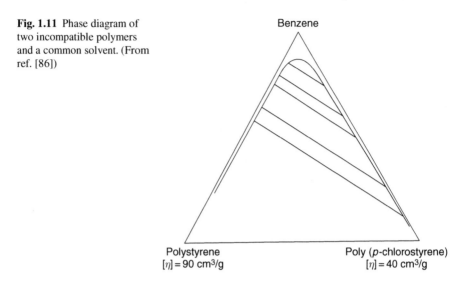

Benzene

Polystyrene
$[\eta] = 90 \text{ cm}^3/\text{g}$

Poly (p-chlorostyrene)
$[\eta] = 40 \text{ cm}^3/\text{g}$

the solutions contain initially long – chain molecules composed of a large number
of A or B units, respectively, the entropy of mixing per unit weight becomes negligible, while the energy of mixing per unit weight, depending on the number of
contact points between dissimilar chain segments remains nearly the same as for the
low molecular weight analog. The solutions of the chain molecules will, therefore,
resist mixing if contacts between dissimilar chain segments involve an expenditure
of energy.

Kern [86] has shown how sensitive the phase separation phenomenon is to small
changes in the structure of the polymers: Poly(vinyl chloride) solutions, which do
not mix with solutions of poly(methyl methacrylate), were found to be miscible with
solutions of polymers of ethyl, propyl, butyl, and isobutyl methacrylate.

1.5 Polymers in Binary Solvents. Cosolvency Effect: Preferential Adsorption Phenomena

Most of the studies on polymers such as polyamides in solution are limited because
the insolubility of these polymers in common solvents. This fact makes their characterization very difficult [87] The degree of order or crystallinity shown by these
polymers caused by intra- and inter- chain interactions affect their solubility and
their general properties.

In order to illustrate this problem, it is interesting to analize the unsubstituted
polyamides prepared from aliphatic diamines and aromatic diacids or from aliphatic
diamines and aliphatic diacids. They are only soluble in solvents such as sulphuric
acid, trifluoracetic acid, m-cresol, etc.

Table 1.5 shows the solubility of the polyamides taken from ref. [88].

Table 1.5 Solubility of polyamides. (From ref. [88])

Polyamides	POS	POT	POTCl
m-Cresol	+	+	−
Formic acid	−	−	−
Sulphuric Acid	+	+	+
Trifluoroacetic acid	+	−	−
Dimethyl sulphoxide	−	−	−
Dimethylformamide	−	−	−
Tetrachloroethane	−	−	−
Cyclohexane	−	−	−
m-Cresol/cyclohexane	+	+	+

* Solubility was determined at 1% concentration; +, soluble at room temperature; −, insoluble at room temperature

It is known from literature that a non-solvent often increases the dissolving power of a solvent for a given polymer [89]. In order to extend the range of the polymer solubility, it has been considered the cosolvency effect, i.e. the synergism effect either of two non-solvents or a solvent with a non-solvent. One illustration of this behavior is that the mixture m-cresol/cyclohexane is a good solvent for poly(octamethylene sebacamide) (POS), poly(octamethylene terephtalamide) (POT) and poly(octamethylene tetrachloroterephtalamide) (POTCl) [88] shown in Figs. 1.12 and 1.13.

The intrinsic viscosity $[\eta]$ for these polymers was chosen as indicative of the thermodynamical power of the binary solvent.

Figures 1.12 and 1.13 show $[\eta]$ values in m- cresol/cyclohexane mixtures for poly (octamethylene terephthamide (POT) and poly(octamethylene sebacamide) (POS).

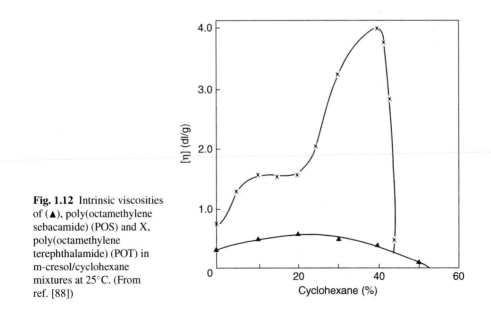

Fig. 1.12 Intrinsic viscosities of (▲), poly(octamethylene sebacamide) (POS) and X, poly(octamethylene terephthalamide) (POT) in m-cresol/cyclohexane mixtures at 25°C. (From ref. [88])

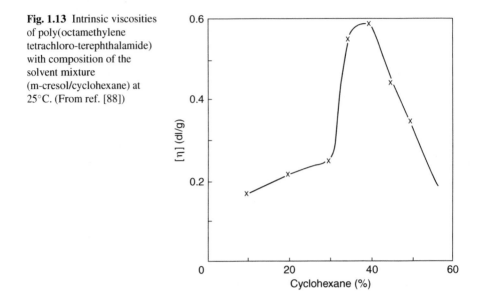

Fig. 1.13 Intrinsic viscosities of poly(octamethylene tetrachloro-terephthalamide) with composition of the solvent mixture (m-cresol/cyclohexane) at 25°C. (From ref. [88])

The intrinsic viscosities of POT and POS in the different solvent/non- solvent mixture are higher than the corresponding intrinsic viscosities in the pure solvent. These positive deviations of [η] relative to the solvent m – cresol suggest that the m – cresol/cyclohexane mixture has a $G^E > O$ [90].

From figure 1.12 it follows that the increase of [η] is much more pronounced in the case of POT than in POS. In both cases, once a particular composition mixture is reached, [η] value decreases to a composition mixture which may be the theta (θ) condition, where phase separation takes place.

Figure 1.13 shows [η] values in m– cresol/cyclohexane mixtures for POTCl. When [η] is plotted against solvent composition for POT polyamide, (Fig. 1.12). This plot shows a maximum at 40 vol% cyclohexane. This feature of attaining a maximum has also been observed in the cosolvent system: POTCl in m – cresol/ cyclohexane at the same solvent composition. (Fig. 1.13).

In general, to explain the observed cosolvent effects, the preferential adsorption phenomena have been invoked. However few topics in the physical chemistry of polymers have evoked so many theories but so little consensus as preferential adsorption. When a polymer is dissolved in a binary solvent mixture, usually one of the solvents preferentially solvates the polymer. This solvent will then be found in a greater proportion in the proximities of the macromolecule with respect to the bulk solution composition. This variation of the solvent composition can cause interesting phenomena such as cosolvency as was discussed before, [11, 91, 92] non – cosolvency [93, 94], and some times variation of the unperturbed polymer dimensions [95, 96]

Preferential adsorption behavior is markedly influenced by various factors [97, 98]. One of them is the chemical structure of the polymer. There are few studies dealing with the effect of the chemical structure of the polymer on preferential adsorption [98,99]. Particularly interesting seems to be the effect of steric hindrance of the side groups. On the other hand, specific interactions are of fundamental importance in the interpretation of preferential adsorption from a thermodynamic point of view. Horta et al. [100] and Katime et al. [98] have studied the effect of specific interactions in the preferential adsorption of polar systems, from theoretical and experimental points of view.

It is well known that the behavior of polymer solutions in binary solvents is also influenced considerably by the thermodynamic nature of the solvents and of the mixture itself [97, 101–103]. The thermodynamic properties of the mixture seems to play an important role in the solvating process of polymeric solutes. Tetrahydrofurane - water (THF/water) is an interesting mixture because THF is one of the most common good solvents for poly(methacrylate)s and water is a non – solvent for these polymers. The mixture is strongly non – ideal. At 298.15 K the excess enthalpy curve is S – shaped, with a zero value near 0.4 mole fraction of water [104]. The excess Gibbs free energy is positive and the excess entropy and volume are negative over the whole mole fraction range [105]. It was reported the enhancement of the polymer solubility when a non – solvent is mixed with a solvent [106]. Particularly this mixture has shown a cosolvency and preferential adsorption effect in poly(4 – tert-butylphenyl methacrylate)(PBPh) and polystyrene (PS). It both systems water is preferentially adsorbed by these polymers [106]. One of the most accepted models to explain the preferential adsorption is that which consider that preferential adsorption is located only in the first solvation layer, but not in the total volume of the coil, and that is uniform along the chain [107]. According with this model the preferential adsorption must be different if the rigidity of the chain and the steric hindrance increase. Gargallo et al. [108] have taken into account the effect of the chemical structure in the preferential adsorption. They have determined the preferential adsorption coefficient λ in poly (phenyl methacrylate) (PPh), poly(4– tert – butylphenyl methacrylate) (PBPh) and poly [4-(1,1,3,3 – tetramethylbutyl methacrylate] (POPh). The preferential adsorption coefficient, λ, was determined from the relation:

$$\lambda = \frac{(\mathrm{dn/dc})_\mu - (\mathrm{dn/dc})_k}{\mathrm{dn/dk}} \tag{1.58}$$

Where $(\mathrm{dn/dc})_k$ is the polymer refractive index increment in the solvent mixture, dn/dk is the variation of the refractive index of the solvent mixture as a function of volumetric composition and $(\mathrm{dn/dc})_\mu$ is the polymer refractive index increment after establishing dialysis equilibrium.

In the case of poly(methacrylates) mentioned above there is a decreasing in the λ values when the aromatic ring has alkyl groups as substituents. This behavior was explained taken into account at least two factors. If the model accepted to explain the preferential adsorption is that which consider that this phenomenon occurs along the polymer chain, the rigidity of the macromolecule must influence the amount of the adsorbed solvent.

This situation can be observed in Table 1.6.

Table 1.6 λ values in the maximum adsorption, rigidity factor σ and the structures of the side groups of poly(methacrylate)s. (From ref. [108]).

Polymer	λ (min) ml g^{-1}	$\sigma = \dfrac{<r_0^2>^{1/2}}{<r_{0f}^2>^{1/2}}$	Side group
PP$_H$	0.62	2.75[a]	
PBP$_H$	0.50	2.90[b]	
POP$_H$	0.44	3.36[c]	

In this Table the rigidity factor, defined by

$$\sigma = <r_0^2>1/2/<r_0^2>1/2 \qquad (1.59)$$

for the different poly (methacrylate)s and the value of the preferential adsorption coefficient λ was compared. Therefore, there is an another factor to take in account, the presence of hydrophobic substituents in the aromatic ring of the polymer chain should provoke a bigger difficulty for the adsorption of water molecules by the polymer. The decreasing of the amount of adsorbed water could be explained also as a consequence of this effect [108].

The preferential adsorption in dilute solution of poly(phenyl methacrylate) and its derivatives in tetrahydrofuran/water mixtures has been studied by differential refractometry and dialysis equilibrium [108]. Relationships were found between the rigidity σ, the inversion composition $v_2(\lambda = 0)$ and the maximum of the adsorption (λ min).

In order to investigate the effect of the ring substitution on preferential adsorption of water in poly(phenyl methacrylate) derivatives, it has been studied the preferential adsorption behavior of poly(2,4 – dimethylphenyl methacrylate) (2,4 – DMP), poly(2,5 – dimethylphenyl methacrylate) (2,5 – DMP) and poly(3,5 – dimethylphenyl methacrylate) (3,5 – DMP) in THF/water mixtures [109].

According to Fig. 1.14, water is preferentially adsorbed at low water composition for the three systems studied. λ also shows a minimum with an inversion point ($\lambda = 0$) at v_2 values with very small difference between the three polymeric systems. These results were compared with other poly (methacrylate)s and poly(styrene)s and they were discussed in terms of the differences in the symmetry of the pendant group [101, 109].

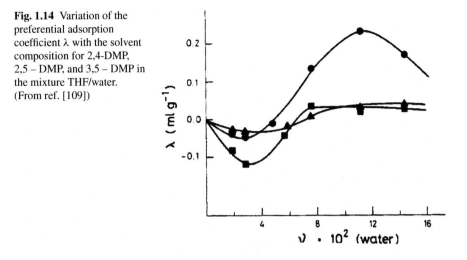

Fig. 1.14 Variation of the preferential adsorption coefficient λ with the solvent composition for 2,4-DMP, 2,5 – DMP, and 3,5 – DMP in the mixture THF/water. (From ref. [109])

The preferential adsorption behavior of poly(vinylpyrrolidone) (PVP) in binary solvent containing aromatic components has been also studied [110]. In this case, it was concerned with the influence of the chemical structure of different binary solvents in the preferential adsorption of this polymer. 2 – propanol – cumene, 2 – propanol – mesitylene, 2 – propanol – p-xylene, 2 – propanol – ethylbenzene and 2 – propanol – toluene. Figure 1.15 shows the variation of λ with the solvent composition. In both cases aromatic components are adsorbed in the range 0 to ≈40%, but the amount of adsorbed molecules is rather different for the two isomers. This result could be explained in terms of steric hindrance due to the isopropyl groups of cumene, which would be reflected in the lower λ value.

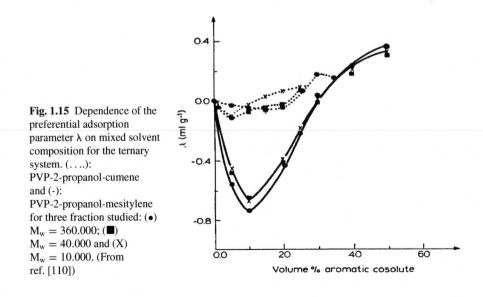

Fig. 1.15 Dependence of the preferential adsorption parameter λ on mixed solvent composition for the ternary system. (. . ..): PVP-2-propanol-cumene and (-): PVP-2-propanol-mesitylene for three fraction studied: (•) $M_w = 360.000$; (■) $M_w = 40.000$ and (X) $M_w = 10.000$. (From ref. [110])

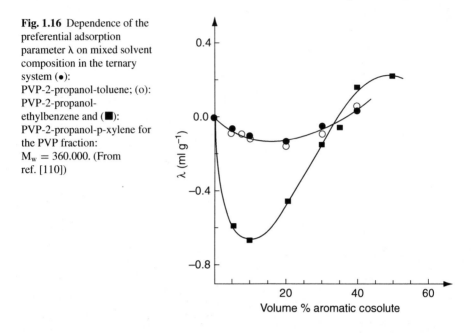

Fig. 1.16 Dependence of the preferential adsorption parameter λ on mixed solvent composition in the ternary system (●): PVP-2-propanol-toluene; (o): PVP-2-propanol-ethylbenzene and (■): PVP-2-propanol-p-xylene for the PVP fraction: $M_w = 360.000$. (From ref. [110])

The variation of the preferential adsorption coefficient λ with the solvent composition for three mixtures is shown in Fig. 1.16. As in the above cases aromatic components are preferentially adsorbed at low compositions, but, p-xylene is more adsorbed than ethylbenzene or toluene.

The results have shown that changes in two parameters which characterize the mixed solvent influence the aromatic compound preferential adsorption by the polymer.

In order to explain the experimental behavior found of λ for PVP in the different mixtures, the polarizability was taken into account because of the methyl groups substituents of the aromatic ring. It is possible to find changes in the nature of the interactions between the polar solute, 2 – propanol, and the aromatic component in the binary mixtures and that these changes affect the λ values. The importance of dipole – induced dipole interactions and steric factors in the formation of a molecular complex between a polar component and a non – polar aromatic solvent has been emphasized on the basis of NMR studies [111, 112]. The molecular interactions in binary liquid mixtures have also been studied on the basis of viscosity measurements. The viscosity data have also been used by Yadava et al. [113, 114] to obtain a value for the interchange energy (W_{visc}) [115] This parameter can be estimated by the equation:

$$\ln \eta_{12} = x_1 \ln \eta_1 + x_2 \ln \eta_2 + x_1 x_2 d_1 \qquad (1.60)$$

suggested by Grunberg [116] In this equation η_1 and η_2 are the viscosities and x_1 and x_2 the mole fractions of the component 1 and 2 respectively. η_{12} is the viscosity

Table 1.7 Values of the preferential adsorption parameter λ in mg of solvent g^{-1} of polymer at the maximum adsorption (minimum in λ), molar volume of the solvent, d parameter and W_{visc} for the aromatic solvents. (From ref. [110])

Aromatic solvent	Molar volume[32] $(cm^3 g^{-1} mol)$	$\lambda\ (mg\,g^{-1})$	d parameter at ~50% mol	W_{visc} Interchange energy (J mol^{-1})
Toluene	106.85	1.12	−0.12	+60.7
Ethylbenzene	123.06	1.22		
p-xylene	123.92	4.76	+0.24	−149.6
Mesitylene	140.00	4.67	+0.20	−134.5

of the solution. The constant d_1 is proportional to W, where W is the interchange energy.

Large negative values d- values for a binary mixtures with cyclohexane indicates the existence of weak interactions due mainly to dispersion forces. Higher, positive d-values for the mixtures prepared with p – xylene and mesitylene seem to be due to enhanced dipole interactions on account of higher polarizability of the methyl substituted benzenes (negative value for W_{visc}).

It has been obtained a surprising result: namely, the λ value is smaller for toluene with a positive W_{visc} than λ values for p – xilene and mesitylene with a negative W_{visc} respectively, as is shown in Table 1.7.

According to Grumberg [116], it appears likely that W_{visc} may be a close approximation to the heat of mixing. Table 1.7 shows that the W_{visc} values are proportional to the λ values. The large negative values of W_{visc} for p – xylene and mesitylene in a polar solvent can be taken as experimental evidence of an effective enthalpic effect in these binary systems relative to toluene – 2 – propanol system, which influence the value of the preferential adsorption coefficient λ.

Table 1.7 shows the values of the preferential adsorption coefficient λ, in mg (mg/g); the molar volumes of the aromatic solvents; d – parameters and the values of the interchange energy (W_{visc}) of the aromatic solvents – 2 – propanol mixtures.

The parameters d, W_{visc}, and λ, have been discussed and correlated well for these systems [110].

1.6 Preferential Adsorption Phenomena: Thermodynamical Description. Association Equilibria Theory

As it was noted before, Preferential or Selective Adsorption is a very common phenomenon in ternary systems composed of a polymer and a binary solvent mixture. There is a great variety of ternary systems that have been studied, mainly those containing at least one polar component [97]. In many cases, specific interactions between polar groups are important, and the formation of hydrogen bonds have to be taken into account. This is the case, especially, when alcohols are components of the systems.

In systems with specific interactions random mixing cannot be assumed. Hence, the thermodynamic theories traditionally used to interpret ternary system properties, such as the Flory – Huggins formalism or the equation of state theory of Flory, are expected not to apply to such systems.

For systems showing strong specific effects such as hydrogen bonding, Pouchly et al. [117] have developed a theoretical framework that is based on the theory of association equilibria [118, 119]. The existence of associated complexes formed by association of individual molecules is explicitly recognized in such theory, and the thermodynamic properties are derived from the equilibrium constants for association. The theory of association equilibria was first fully developed and applied only for the particular case in which one of the two liquids in the mixed solvent self – associates and interacts specifically with the polymer while the other solvent is inert [119]. Horta, Radić and Gargallo [120] have extended the theoretical treatment to the case in which one of the liquids self – associates, interacts specifically with the polymer, and also interacts specifically with the molecules of the other liquid. This case is more general, and the theoretical results obtained for it include the previous case of one liquid being inert as a particular case [120]. This last theoretical treatment was used to derive the amount of preferential adsorption for polymer – mixed solvent systems in which one of the two liquids in the mixed solvent (B) autoassociates and interacts specifically with the polymer and with the other liquid (A). The model takes into account the constants for the association of one molecule of B to one site in A and of one molecule of B to one site in the polymer and also the corresponding constants for the multiple self - association of B for the case where the first B molecule is associated either with A or with the polymer or is free.

The theoretical results were applied to the systems poly(alkyl methacrylate)s. Alkyl = Me, Et, iBu in the mixed solvent methanol (B) + 1,4-dioxane (A) [98].

Figure 1.17 shows the comparison of theory and experiment for preferential adsorption coefficient λ of poly(alkyl methacrylate)s in 1,4-dioxane-methanol.

A quantitative agreement was found between theory and the experimental results dealing with the dependence of the preferential adsorption coefficient λ with the composition, in the cases that were chosen as examples: Poly(alkyl methacrylates)/1, 4 – dioxane/methanol [120].

On the other hand, in this kind of systems, the experimental results have indicated that the preferential adsorption is markedly influenced by the size of the side groups attached to the polymer backbone, [120] as has been reported for other related ternary systems [109, 121].

In fact, the adsorption of methanol diminishes as the size of the side group increases and finally disappears when the lateral group is bulky enough.

It was also reported experimental results about the preferential adsorption of a family of poly(dialkyl itaconates) [poly[1 – (alkoxycarbonyl) – 1 – [(alkoxycarbonyl) – methyl] ethylene] in 1,4 – dioxane/methanol. The chemical structures of the polymers are shown Fig. 1.18 [122].

The dependence of the preferential adsorption coefficient λ, for PDMI, PDEI, PDPI, and PDBI in 1, 4-dioxane (A)/methanol (B) mixtures, as a function of

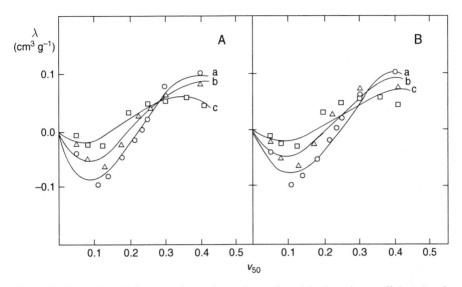

Fig. 1.17 Comparison of theory and experiment for preferential adsorption coefficient, λ, of poly(alkyl methacrylate)s in 1,4-dioxane-methanol. (μ_{BO} = methanol volume fraction). Points: Experimental results from ref. [6]. (○) PMMA (alkyl = Me); (Δ) PEMA (Et); (□) PiBM (iBu). Association equilibria theory. (2-A) Calculated with the parameter values shown in Table I and numbered as 4–6, Curves: (a) PMMA; (b) PEMA, (c) PiBMA. (2-B) Calculated with the parameter values shown in Table I and numbered as 10–12. Curves: (a) PMMA; (b) PEMMA; (c) PiBMa. (From ref. [120])

the methanol fraction in the mixed solvent, $u_{B0}(u_{A0} = 1 - u_{BO})$, is shown in Fig. 1.19.

As can be seen, at low u_{B0}, methanol is preferentially adsorbed by the polymer, and also in all the cases there is an inversion in solvation ($\lambda = 0$).

A quantitative description of the variation of λ and its dependence with the nature of the side groups of the polymer in these systems was found by applying the association equilibria theory.

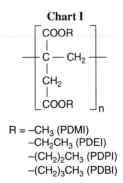

Chart I

Fig. 1.18 Poly(itaconates) as compared to only one carbonyl in poly-(methacrylates). (From ref. [122])

$R = -CH_3$ (PDMI)
$-CH_2CH_3$ (PDEI)
$-(CH_2)_2CH_3$ (PDPI)
$-(CH_2)_3CH_3$ (PDBI)

Fig. 1.19 Variation of
preferential adsorption
coefficient, λ, as function of
methanol volume fraction
μ_{BO}, for PDMI ($\circ$), PDEI
($\triangle$), PDPI ($\square$), and PDBI ($\bullet$),
at 298 K. (From ref. [122])

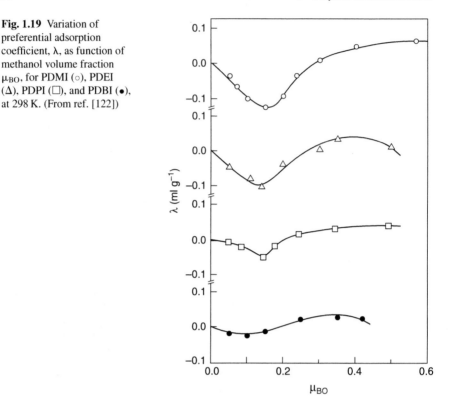

Figure 1.20 shows the results of λ calculated with the association equilibria theory using the parameter values derived from the theory. They give a very good description of the experimental results [122].

Most of the unique properties exhibited by polymers are a result of the quasi-unlimited number of spatial arrangements that long chains can assume [123]. The knowledge of the energetic effects accompanying the changes from one configuration to another, in correlation with the structural characteristics of the chains, is paramount to interpret and predict the physical properties of polymeric materials [124].

The thermodynamic state of a polymer-solvent system is completely determined, as it was analized before, at fixed temperature and pressure by means of the interaction parameter g. This g is defined through the noncombinatorial part of the Gibbs mixing function, ΔG_M. The more usual interaction parameter, χ, is defined similarly but through the solvent chemical potential, $\Delta\mu_1$, derived from ΔG_M.

In ternary systems composed of one polymer and two liquids or of two polymers and one solvent, the total Gibbs mixing function of the system can be written in terms of the g interaction parameters of the corresponding binary pairs, according to the Flory – Huggins formalism [11]. When studying polymers in mixed solvents, it has been customary to introduce an additional interaction parameter, called ternary,

Fig. 1.20 Comparison of theory and experiment for the preferential adsorption coefficient, λ, of poly(dialkyl itaconates) in 1,4-dioxane/methanol (μ_{BO} = methanol volume fraction). Points: experimental results for PDMI ($\circ$), (alkyl = Me); PDEI ($\triangle$) (Et); PDPI ($\square$) (nPr); PDBI ($\bullet$) (nBu). Curves: Association equilibria theory calculated with the set of parameter values shown in Table II. (From ref. [122])

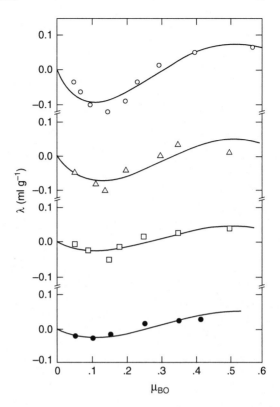

[125] g_T. It has been the object of numerous investigations to attempt the prediction of the total sorption and the preferential adsorption of polymers in mixed solvents from the interaction parameters of the binary pairs and use the ternary parameter as adjustable. This has shown that such a ternary parameter is of great importance [97].

In the same way that in a binary polymer–solvent system the interaction parameter g constitutes the complete thermodynamic description of the system and predicting g is the test for any theoretical model, so in a ternary system the ternary parameter can be viewed as describing the characteristics of the system and being the goal for any interpretation of it.

However, a correct unequivocal determination of the ternary parameter has not been possible up to now, due to the lack of g data for the binary systems.

The case of infinite dilution of the polymer is the one for which total and preferential sorption have been most extensively studied. In this dilute solution limit the interpretation of the preferential adsorption coefficient, λ, requires knowledge of the g interpretation parameters at infinite dilution, g^0, for the polymer in each one of the pure liquids [9].

The lack of knowledge of g^0 parameter values has led to different approximations, the crudest of them [125] being to substitute these parameters for their corresponding χ_s^0, which implies the assumption that polymer – solvent interaction parameters are not dependent on polymer concentration, in clear contradiction with overwhelming experimental evidence on the contrary [126]. To avoid such an approximation, it has been proposed to use the difference $g_{13}° - lg_{23}°$ as adjustable from the preferential adsorption data [9,127,128] jointly with the ternary parameter g_T.

However, for those binary polymer – solvent systems in which data of χ as a function of concentration are available, it is possible to obtain the g interaction parameters directly. With the g's thus determined, it should be unnecessary to use any approximation or adjustment of binary parameters in the study of ternary systems. To provide values of g^0 of binary polymer – solvent systems, Masegosa et al. [6] have calculated g^0 for 41 polymer – solvent systems for which data of χ vs. concentration were available in the literature. The values of g^0 calculated are collected in Table 1.2.

These values of g^0 have been compared with those calculated from Flory theory.

References

1. D. H. Whiffen, "Manual of Symbols and Terminology for Physicochemical Quantities and units, Appendix I: Solutions". Pure Appl. Chem., 51, 1 (1979).
2. G. Jancso, D. V. Fenby, "Thermodynamics of Dilute Solutions". J. Chem. Educ., 60, 382 (1983)
3. P. J. Flory, J. Chem. Phys. (1941) 10, 51 (1952).
4. M. L. Huggins, Ann. N.Y. Acad. Sci., 43, 1 (1942).
5. G. Jannink, J. des Cloizeaux, J. Phys.: Condens. Matter., 2, 1 (1990).
6. R. M. Masegosa, M. G. Prolongo, A. Horta, Macromolecules, 19, 1478 (1986)
7. P. J. Flory, J. Am. Chem. Soc., 87, 1833 (1965)
8. A. Horta, Macromolecules, 18, 2498 (1985)
9. J. Pouchly, A. Zivny, Makromol. Chem., 183, 3019 (1982)
10. J. H. Hildebrand, J. M. Prausnitz, R. L. Scott "Regular and Related Solutions" Van Nostrand Reinhold Ltd. 1970, Chapter 10
11. P. J. Flory, "Principles of Polymer Chemistry" Cornell University Press, Ithaca, New York, 1953.
12. O. Fuchs, Fortschr. Chem. Forsch., 11, 74 (1968)
13. K. C. James, Edu. Chem., 9, 220 (1972)
14. A. F. M. Barton, "Handbook of Solubility Parameters and other Cohesion Parameters" CRC Press, Boca Raton/Florida (1983)
15. M. R. J. Dack, "The Importance of Solvent Internal Pressure and Cohesion to Solution Phenomena", Chem. Soc. Rev. (London), 4, 211 (1975)
16. A. Horta, I. Fernandez – Pierola, Macromolecules, 14, 1519 (1981)
17. H. Tompa,"Polymer Solutions" Academic Press, New York (1965).
18. H. A. Stuart, Die Physik der Hochpolymerem, vol 2 (Springer, Berlin, 1952)
19. H. Morawetz, "Macromolecules in Solution", Interscience New York (1965).
20. I. Dort, Polymer, 29, 490 (1988)
21. H. J. Hildebrand, R. L. Scott "The solubility of Nonelectrolytes", Am. Chem. Soc. Monograph No 17, Reinhold, New York, Chs. 20 (1949)
22. M. L. Huggins, J. Chem. Phys., 9, 440 (1941)
23. M. L. Huggins, J. Phys. Chem., 46, 151 (1942)
24. P. J. Flory, J. Chem. Phys., 9, 660 (1941)
25. P. J. Flory, J. Chem. Phys., 10, 51 (1942)
26. R. F. Weimer, J. M. Prausnitz, Hydrocarbon Process. Petrol. Refiner 44, 237 (1964)

27. C. M. Hansen, J. Paint Technol., 39, 104 (1967)
28. D. M. Koenhen, C. A. Smolders, J. Appl. Polym. Sci., 19, 1163 (1975)
29. R. Sattelmeyer, Adhesion, 20, 278 (1976)
30. R. A. Orwoll, Rubber Chem. Technol., 50, 451 (1977)
31. H. Yamakawa, "Modern Theory of Polymer Solution", Harper and Row, New York (1971)
32. A. K. Kashyap, U. Kalpagam, C. Reddy, Polymer, 18, 878 (1977)
33. J. Velikovic', J. Filipovic', S. Co'seva. Eur. Polym. J., 15, 521 (1979)
34. M. Maillols, L. Bardet, S. Gromb, Eur. Polym. J., 14, 1015 (1978)
35. G. M. Bristow, W. F. Watson, Trans. Far. Soc., 54, 1731 (1958)
36. K. Ito, J. E. Guillet, Macromolecules, 12, 1163 (1979)
37. C. M. Kok, Alfred Rudin, J. Coating Technol., 55, 57 (1983)
38. H. L. Frish, R. N. Simha, "Rheology. Theory and Applications", Vol. 1, Academic Press, New York, (1956)
39. M. Yazdani, L. Gargallo, D. Radić, Eur. Polym. J., 21, 461 (1985)
40. M. Bohdanecký, Collect. Czech. Chem. Commun., 35, 1972 (1970)
41. K. K. Chee, J. Appl. Polym. Sci., 23, 1639 (1979)
42. M. Bohdanecký, Z. Tuzar, Collect. Czech. Chem. Commun., 34, 3318 (1969)
43. J. Moacanin, J. Appl. Polym. Sci., 1, 272 (1959)
44. M. Becerra, D. Radić, L. Gargallo, Makromol. Chem., 179, 2241 (1978)
45. H. Fugita, "Polymer Solutions", Elsevier, Tokyo, (1990)
46. J. G. Kirdwood, J. Riseman. J. Chem. Phys., 16, 565 (1948). GKirdwood
47. P. J. Flory, J. Chem. Phys., 17, 303 (1949)
48. W. G. McMillan, J. Shimada, J. Chem. Phys., 13, 276 (1945)
49. N. Saito, "Kobunshi Butsuri (Polymer Physics)". (Rev. Ed.), Syokabo, Tokyo, Chapter 4, (1967)
50. T. Ojeda, D. Radić, L. Gargallo, D. Boys, Macromol. Chem., 181, 2237 (1980)
51. W. R. Krigbaum, P. J. Flory, J. Polym. Sci., 9, 503 (1952)
52. T. G. Fox, P. J. Flory, A. M. Bueche, J. Am. Chem. Soc., 73, 285 (1951)
53. G. C. Berry, E. F. Casassa, J. Polym. Sci., Part D, 4, 1 (1970)
54. M. Daoud, J. P. Cotton, B. Farnoux, G. Jannink, G. Serma, H. Benoit, R. Duplessix, C. Picot, P.-G de Gennes, Macromolecules, 8, 804 (1975)
55. J. S. King, W. Boyer, G. D. Wignall, R. Ullman, Macromolecules, 18, 709 (1985)
56. B. L. Hager, G. C. Berry, J. Polym. Sci., Polym. Phys. Ed., 20, 911 (1982)
57. W. W. Graessley, Polymer, 21, 258 (1980)
58. Q. Ying, B. Chu, Macromolecules, 20, 362 (1987)
59. D. W. Schaefer, C. C. Han, in "Dynamic Light Scattering", Ed. R. Pecora, Plenum, New York, Chapter 5 (1985)
60. M. Doi, S. F. Edwards, "The Theory of Polymer Dynamics", Oxford: Clarendon (1986).
61. J. des Cloizeaux, G. Jannink, Les Polymeres en Solution; leur Modelisation et leur Structure (Les Ulis: Editions de Physique)
62. G. Allegra, F. Ganazzoli, G. Raos, TRIP, 7, 293 (1996)
63. W. H. Stockmayer, Makromol. Chem., 35, 54 (1960)
64. H. Yamakawa, Macromolecules, 26, 5061 (1993)
65. A. Y. Grosberg, D. V. Kuznetsov, Macromolecules, 26, 4249 (1993)
66. I. H. Park, Q. W. Wang, B. Chu, Macromolecules, 20, 1965 (1987)
67. E. A. Di Marzio, Macromolecules, 17, 969 (1984)
68. I. C. Sanchez, Macromolecules, 12, 276 (1979)
69. B. Chu, I. H. Park, Q. W. Wang, C. Wu, Macromolecules, 20, 2883 (1987)
70. B. Chu, J. Wu, Z. L. Wang, Prog. Colloid Polym. Sci., 91, 142 (1993)
71. H. S. Chan, K. A. Dill, Phys. Today, 46, 24 (1993)
72. C. B. Post, B. H. Zimm, Biopolymers, 21, 2139 (1982)
73. P. J. De Gennes, "Scaling Concepts in Polymer Physics", Cornell University Press Ltd. Ithaca and London. (1979)
74. I. C. Sanchez, J. Phys. Chem., 93, 6983 (1989)
75. B. Widom, Physica A 194, 532 (1993)

76. P. Debye, J. Chem. Phys., 31, 680 (1959)
77. P. Debye, H. Coll, D. Woermann, J. Chem. Phys., 32, 939 (1960)
78. B. Chu, Phys. Lett. A, 28, 654 (1969)
79. P. G. de Gennes, Phys. Lett. A, 26, 313, (1968)
80. A. Vrij, M. W. J. Van Den Esker, J. Chem. Soc. Faraday Trans., 268, 513 (1972)
81. G. Raos, G. Allegra J. Chem. Phys., 104, 1626 (1996)
82. G. B. Fields, D. O. V. Alonso, D. Stigter, K. A. Dill, J. Phys. Chem., 96, 3974 (1992)
83. L. Gargallo, D. Radić, F. Martinez-Piña, Eur. Polym. J., 33, 1767 (1997)
84. D. Radić, L. Gargallo, Thermochimica Acta, 180, 241 (1991)
85. M. Urzúa, L. Gargallo, D. Radić, J. Macromol. Sci. Phys. B39, 143 (2000)
86. R. J. Kern, J. Polym. Sci., 33, 524 (1958)
87. P. W. Morgan, S. L. Kwolek, Macromolecules, 8, 104 (1975)
88. D. Radić, D. Boys, L. Gargallo, Polymer, 18, 121 (1977).
89. K. Sarkar, S. R. Palit, J. Polym. Sci. (C), 30, 69 (1970).
90. A. P. Dondos, D. Patterson, J. Polym. Sci. (A-2), 7, 289 (1969)
91. J. M. G. Cowie, J. Polym. Sci. C23, 267 (1968)
92. H. Maillols, L. Bardet, S. Gromb, Eur. Polym. J., 14, 1015 (1978)
93. B. A. Wolf, M. M. Willms, Makromol. Chem., 179, 2665 (1978)
94. O. Fuchs, Kunstoffe 43, 409 (1953)
95. A. Dondos, P. Rempp, H. Benoit, J. Polym. Sci., 30, 9 (1970)
96. L. Gargallo, Makromol. Chem., 177, 233 (1976)
97. L. Gargallo, D. Radić, Adv. Colloid Interface Sci., 21, 1 (1984)
98. I. Katime, L. Gargallo, D. Radić, A. Horta, Makromolek. Chem., 186, 2125 (1985)
99. A. Dondos, H. Benoit, Makromolek. Chem., 133, 119 (1970)
100. A. Horta, D. Radić, L. Gargallo, I. Katime, Makromolek. Chem., 189, (1988)
101. D. Radić, L.Gargallo, Eur. Polym. J., 18, 151 (1982)
102. P. Munk, M. T. Abijaoude, M. E. Halbrook, J. Polym. Sci. Polym. Ohys., 16, 105 (1978)
103. L.Gargallo, D. Radić, I. Katime, Eur. Polym. J., 16, 383 (1980)
104. H. Nakayama, K. Shinoda, J. Chem. Thermodyn., 3, 401 (1971)
105. K. W. Morcom, R. W. Smith, Trans. Faraday Soc., 66, 1073 (1970)
106. D. Radić, L. Gargallo, Polymer, 22, 1045 (1981)
107. L. Gargallo, D. Radić, I. Fernandez-Pierola, Makromol. Chem., Rapid Commun., 3, 409
     (1982)
108. L. Gargallo, H. Rios, D. Radić, Polymer Bull., 11, 525 (1984)
109. D. Radić, N. Hamidi, L. Gargallo, Eur. Polym. J., 24, 799 (1988)
110. L. Gargallo, D. Radić, Polym. Commun., 26, 149 (1985)
111. J. Homer, R. R. Yadava, Tetrahedron, 29, 3853 (1973)
112. R. R. Yadava, S. S. Yadava, Indian J. Chem., 16 A, 826 (1978)
113. J. Homer, R. R. Yadava, J. Chem. Soc., Faraday Trans., 1, 611 (1974)
114. R. R. Yadava, S. S. Yadava, Indian J. Chem., 18A, 120 (1979)
115. R. J. Fort, W. R. Moore, Trans. Faraday Soc., 62, 1112 (1966)
116. L. Grumberg, Trans. Faraday Soc., 50, 1293 (1954)
117. J. Pouchly, A. Zivny, Makromol. Chem., 186, 37 (1985)
118. J. Pouchly, A. Zivny, K. Solc, Collect. Czech. Commun., 37, 988 (1972)
119. J. Pouchly, Collect. Czech. Chem. Commun., 34, 1236 (1969)
120. A. Horta, D. Radić, L. Gargallo, Macromolecules, 22, 4267 (1989)
121. L. Gargallo, N. Hamidi, I. Katime, D. Radić, Polym. Bull., 14, 393 (1985)
122. A. Horta, L. Gargallo, D. Radić, Macromolecules, 23, 5320 (1990)
123. P. J. Flory, "Statistical Mechanics of Chain Molecues, Interscience", New York (1969)
124. E. Riande, Macromol. Chem. Macromol. Symp., 2, 179 (1986)
125. B. E. Read, Trans. Faraday Soc., 56, 382 (1960)
126. P. J. Flory, J. Chem. Soc. Faraday Dis., 49, 7 (1970)
127. S. G. Chu, P. Munk, Macromolecules, 11, 879 (1978)
128. J. E. Figueruelo, B. Celda, A. Campos, Macromolecules, 18, 7511 (1985)

Chapter 2
Viscoelastic Behaviour of Polymers

Summary In this chapter, a discussion of the viscoelastic properties of selected polymeric materials is performed. The basic concepts of viscoelasticity, dealing with the fact that polymers above glass-transition temperature exhibit high entropic elasticity, are described at beginner level. The analysis of stress-strain for some polymeric materials is shortly described. Dielectric and dynamic mechanical behavior of aliphatic, cyclic saturated and aromatic substituted poly(methacrylate)s is well explained. An interesting approach of the relaxational processes is presented under the experience of the authors in these polymeric systems. The viscoelastic behavior of poly(itaconate)s with mono- and disubstitutions and the effect of the substituents and the functional groups is extensively discussed. The behavior of viscoelastic behavior of different poly(thiocarbonate)s is also analyzed.

Keywords Viscoelasticity · Glass transition temperature · Relaxational processes · Dielectric behavior · Dynamic mechanical behavior · Poly(methacrylate)s · Poly (itaconate)s · Poly(thiocarbonate)s · Spacer groups · Side chains · Molecular motions

2.1 Introduction

Viscoelastic materials are those which exhibit both viscous and elastic characterists. Viscoelasticity is also known as anelasticity, which is present in systems when undergoing deformation. Viscous materials, like honey, polymer melt etc, resist shear flow (shear flow is in a solid body, the gradient of a shear stress force through the body) and strain, i.e. the deformation of materials caused by stress, is linearly with time when a stress is applied [1–4]. Shear stress is a stress state where the stress is parallel or tangencial to a face of the material, as opposed to normal stress when the stress is perpendicular to the face. The variable used to denote shear stress is τ which is defined as:

$$\tau = \frac{VQ}{It} \tag{2.1}$$

L. Gargallo, D. Radić, *Physicochemical Behavior and Supramolecular Organization of Polymers*, DOI 10.1007/978-1-4020-9372-2_2,
© Springer Science+Business Media B.V. 2009

for shear stress in a beam where V is the shear force, Q is the first moment of area, t, is the thickness in the material perpendicular to the shear and I is the second moment of area of the cross section [1–4].

When an elastic material is stressed, there is an immediate strain response. The classical representation of stress strain response in a perfectly elastic material is schematically represented in Fig. 2.1.

As can be observed when the stress is removed the strain also returns to zero. In a perfectly elastic material all the deformation is returned to the origin [3, 4]. If this energy is not stored elastically then it would be dissipated as either heat or sound. Tyre suqueal and the heat build-up in the sidewalls of care tyres are good examples of such dissipation [2–4]. Another example is when a plastic beaker struck, emits a dull note of short duration, which is quite different from the ringing note emitted by a bell or a crystal wine glass [2, 3]. This property of high mechanical damping is another manifestation of viscoelasticity. It is a property that is frequently of value, for instance in shock absorbers as McCrum et al pointed out [1]. In plastic structures subject to forced oscillation, mechanical vibrations at the natural frequencies of the structure do not easily build up due to the high damping capacity of the plastic. According to McCrum [1] a good example of this is the application of plastic materials in sailing craft, particularly in hull construction. Vibration of the hull stimulated by elements are rapidly damped.

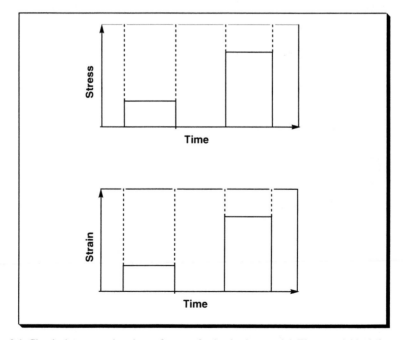

Fig. 2.1 Classical stress strain scheme for a perfectly elastic material. The material is deformed in a proportion to the applied stress and returns to its original state when the stress is released

If the material is linear and elastic then the applied stress σ is directly proportional to the strain ε. Then for simple tension:

$$\sigma = E\varepsilon \tag{2.2}$$

where σ is the stress, E is the elastic modulus of the material, known as Young's modulus, an is considered to be a property of the material, and ε is the stress that occur under the given stress, similar to Hooke's law. Polymers are the most important viscoelastic systems. The typical experiment consider that if a weight is hanging on a polymeric sample, the result is a lengthening of the polymer which produce a difference on the length of the sample (ΔL). Therefore, the stress, σ, can be defined as strength by unit of area,

$$\sigma = \frac{F}{A} \tag{2.3}$$

and the strain of the sample ε, as lengthening by unity of original length

$$\varepsilon = \frac{\Delta L}{L} \tag{2.4}$$

The mechanical characteristic of the material can be measured through the elastic modulus, E, and the docility, D, through equation (2.5).

The viscous components can be modeled as dashpots such that the stress–strain rate relationships which can be given as:

$$\varepsilon = D\sigma \tag{2.5}$$

According with their definition the elastic modulus take into account the stress necessary to produce an unitary lengthening and the docility the strain produced due to the application of a unity stress [1–7].

As it was mentioned above mechanical properties of polymers are strongly dependent on the temperature. Therefore, E and D, for a polymeric sample are dependent on the temperature at which the experiment is performed. On the other hand the mechanical properties of polymers are also dependent on time. Therefore E and D are not constant at one temperature but evolve with time i.e. E(t), D(t) [7].

The complex relationship between the configurational distorsion produced by a perturbation field in polymers and the Brownian motion that relaxes that distorsion make it difficult to establish stress–strain relationships. In fact, the stress at that point in the system depends not only on the actual deformation but also on the previous history of the deformation of the material.

One important characteristics of polymeric materials is their viscoelastic behavior. This means that polymer is elastic because after a strain due to the application of a stress, it is capable to recovers. On the other hand, polymers are viscous because their capability to creep after the strain. How can be explained this peculiar behavior which is observed on polymeric materials?. The application of a weight on a poly-meric film or thread (fiber) produce a strain which not be constant but will have a progressive variation on time. On release of the stress a molecular reorganization take place and molecules slowly recover their former spatial arrangements and the strain simultaneously return to zero. Due to the strain involve all the molecular segments of the polymeric chain, the main influence on this behavior is the con-catenation of the monomer units, what is known as "polymer effect" [1, 2].

It is important to quantify the dynamic viscoelastic properties of the materials. Normally the analysis of these systems is performed using the frequency as the variable, and the relationship between the dynamic parameters and the parameters for step-function suppose the application of an oscillatory shear strain with angular frequency ω expressed as:

$$\gamma = \gamma_0 \sin \omega t \tag{2.6}$$

Due to the viscoelastic nature of the material the stress response, after the appli-cation of the oscillatory shear strain, is also a sinusoidal but out of phase relative to the strain what can be represented by equation (2.7) as

$$\sigma = \sigma_0 \sin(\omega t + \delta) \tag{2.7}$$

where δ is the lag in the phase angle. This behavior can be observed by using the classical vector representation of an alternating stress leading and alternating strain by phase angle δ on Fig. 2.2.

The classical procedure for the analysis of stress strain parameters start from the expansion of σ from equation (2.7) as follows:

$$\sigma = (\sigma_0 \cos \delta) \sin \omega t + (\sigma_0 \sin \delta) \cos \omega t \tag{2.8}$$

therefore the stress consists of two components: one in phase with the strain ($\sigma_0 \cos \delta$); the other 90° out of phase ($\sigma_0 \sin \delta$).

The relationship between stress and strain in this dynamic case can be defined by writing

$$\sigma = \gamma_0[G' \sin \omega t + G'' \cos \omega t] \tag{2.9}$$

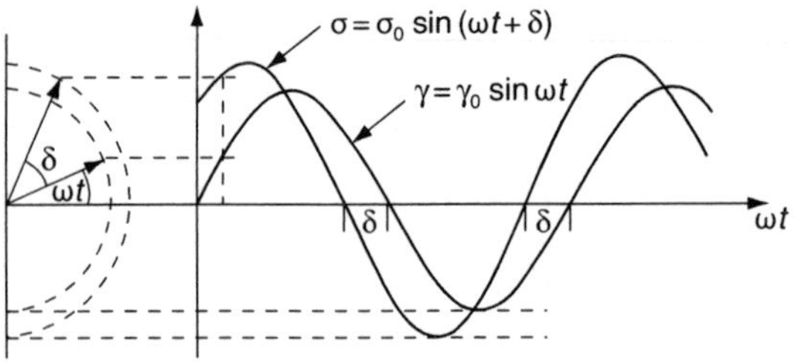

Fig. 2.2 Vector representation of an alternating stress leading and alternating strain by phase angle δ. (From ref. [1])

where

$$G' = \frac{\sigma_0}{\gamma_0} \cos \delta \qquad (2.10)$$

and

$$G'' = \frac{\sigma_0}{\gamma_0} \sin \delta \qquad (2.11)$$

The ratio G''/G' is the tangent of the of the phase angle δ which is defined as:

$$\tan \delta = \frac{G''}{G'} \qquad (2.12)$$

This is an important parameter to analyze the viscoelastic behavior of different materials mainly in the case of polymeric materials where the dependence of tan δ with the chemical structure of the polymeric materials give important information about the relaxation processess that take place is these systems. tanδ is commonly used as a first experimental approach to obtain information about the viscoelastic behavior of polymers as function of the frequency, where it is possible to reach experimental information about the effect of the side chain structure of the polymers on conformational and relaxational responses.

2.2 The Nature of Viscoelasticity

Polymers are viscoelestic at all temperatures. Nevertheless they are not simple elastic solids and the effect of temperature on the response of viscoelastic systems to the perturbation show a very interesting trend. At $T = Tg$ the strain against time remain approximately constant. As the temperature increases the variation of strain with time increases in a square root shape which is larger the higher temperature. See Fig. 2.3.

According to results reported in the literature [1–13] if the shear stress is canceled out after steady-state conditions are reached, the time dependence of the recoverable deformation $[\varepsilon_r(t) - \sigma\varepsilon(t)/\eta]$ is obtained where $\varepsilon(t)$ is the shear strain, σ is the stress and η is the viscosity (2.6). The higher temperature, the greater the unrecoverable contribution to the shear deformation i.e. the viscous deformation. Figure 2.3 shows the effect of temperature on the strain.

The state that it is possible to find a polymeric material is strongly dependent on the chemical structure. In fact, depending on the nature of the chemical functional groups inserted in the polymeric chain, is the thermal, mechanical and dielectric behavior. Depending of the chemical structure of the main chain as well as of the polymeric material is the general behavior of the polymer. Nevertheless, a polymer can be change from one state to other with the variation of the temperature [1–14]. This is one of the most important thermodynamic variables to be taken into account

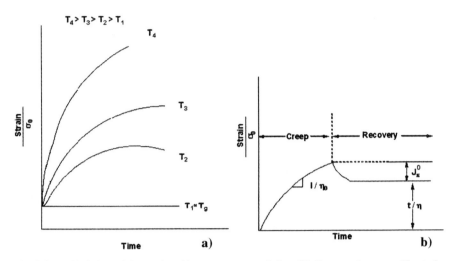

Fig. 2.3 (a) Variation of the strain with temperature and time.(b) Creep and recoverable strain. (From ref. [5])

in the characterization and application of a polymeric material. As a general procedure, polymers are characterized, in the solid state, by their thermal properties, such as melting temperature, crystallization temperature, in the case of semi-crystalline polymers, glass transition temperature (Tg), which is one of the most important characteristic of the polymer due to the general amorphous character of the polymeric materials. Another structural aspect that should be considered in the general behavior of polymers, are the molecular weight, the crystallinity degree, and the branching degree of the polymer [1–7, 15–20]. The majority of the polymeric materials used as industrial ones are obtained by radical polymerization, therefore, this kind of materials are not only amorphous, but they have a wide molecular weight distribution, what is determinant in the behavior of the polymer from different points of view [7, 14]. For instance, glass transition temperature (Tg) is dependent on the molecular weight of the polymer, mainly for low molecular weights [15–18]. On the other hand the degree of crystallinity of the polymers, and the degree of branching could be the responsible of the thermal behavior of the polymeric materials. As a general comment, the mechanical behavior of polymers is hardly dependent on the temperature [1–13, 21, 22].

Above the glass transition temperature, the response of these materials to a mechanical perturbation field involves several types of molecular motions. For example the rearrangement of flexible chains may be very fast on the length scale of repeating unit [5]. These motions imply some type of cooperativity in the conformational transitions that produce them. Cooperativity occurs even as the relaxation propagates along the chains, involving a growing number of segments of the backbone as time passes. At very long times disentanglement of the chain takes place, and the longest relaxation time associated with this process shows a strong dependence on both the molecular weight and the molecular architecture of the system. The disentanglement process governs the flow of the system. As a consequence of the complexity of the molecular responses, polymer chains exhibit a wide distribution of relaxation times that extend over several decades in the time or frequency domains. At short times the response is mainly elastic, whereas at long times it is mainly viscous. The elastic component of the deformation is of an entropic nature, and consequently it is time dependent [5–9].

2.3 Molecular Theory

Polymers, because of their long chemical structure cohere as solids even when discrete section of the chain are undergoing Brownian motions moving by diffusional jump processes from place to place. This is the main difference between elastic solids and polymers [1–7].

Elastic materials strain instantaneously when stretched and just as quickly return to their original state once the stress is removed. Viscoelastic materials have elements of both of these properties and, as such, exhibit time dependent strain. Whereas elasticity is usually the result of bond stretching along crystallographic planes in an ordered solid, viscoelasticity is the result of the diffusion of atoms or molecules inside of an amorphous material [3–6].

The viscoelastic properties are highly temperature-dependent so that the maximum temperature should be always specified and taken into account. Polymers at room temperature behave by different ways i.e. hard solids, elastic liquids, rubbers, etc [1,7].

In order to know how is the variation of the mechanical properties of the polymers with temperature it is necessary to know the time of the measurements. In fact, E and D values obtained at different temperatures are comparable themselves if the time considered for the experiment is the same. Therefore the comparison of the experiments at different temperatures at the same time are isochrones [1–7, 15–20].

It is interesting to analyze the effect of the temperature on the elastic modulus. The classical schematic representation of this behaviour is shown on Fig. 2.4:

At low temperatures (A zone) the polymer is found in the vitreous state. In this state the polymer behave as a rigid solid with low capacity of motions and then the "strain" is very low. To produce a small strain it is necessary a great stress. Therefore in this zone only specific and local motions take place and the polymer can be considered as undeformable. As the temperature increases (B zone) the glass transition temperature, Tg, is reached and the motions of the different parts of the polymers increases but is not enough to produce important strain. Under this conditions the polymers behave as a rubber. If the temperature remain increasing (C zone) the polymer behave as deformable and elastic rubber but the modulus is small. In this zone the motions of the side chains and also of the main chain increases due to the application of the strain.

If the amorphous polymer is not submitted to a strain the macromolecule take a random coil conformation but when is submitted to a strain a rearrangement take place and, a new structural order of the macromolecules with a stretch conformation

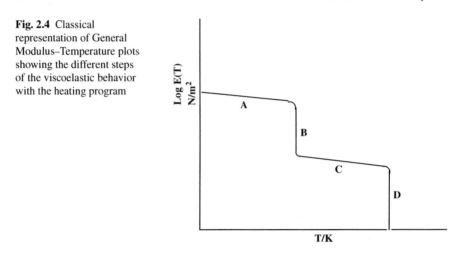

Fig. 2.4 Classical representation of General Modulus–Temperature plots showing the different steps of the viscoelastic behavior with the heating program

is observed. These changes are strongly dependent on the nature of the functional groups on the polymer. These functional groups are able to produce different effects on the response of the polymer to different field force [21, 22].

Finally when the temperature increases several degree over the Tg (D zone) the mobility of the molecules is larger and the application of the force produce a creep of the chains relative to the other chains. This is the general behaviour of linear amorphous polymers.

In the case of polymer networks there is not possibility of viscous flow irrespective of the increment on the temperature. The network when reach a temperature higher than Tg, the modulus is like that for a rubber with a constant value (see curves 3 and 4) in Fig. 2.5.

For linear polymers the elastic behavior is explained by the coil shape which behave for this purpose similar to an entanglement. These entanglements acts as

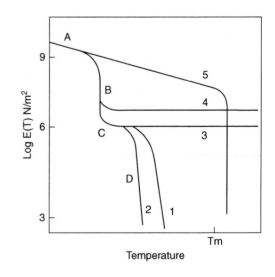

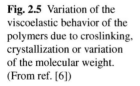

Fig. 2.5 Variation of the viscoelastic behavior of the polymers due to croslinking, crystallization or variation of the molecular weight. (From ref. [6])

knots points in the net, these are not formal knots, but entanglements and strong interaction among the chains take place. In the case of elastomeric polymers there is no time dependence due to the fact that in these materials the interbreed have permanent bonds. When the stress is applied the elastomer reach the equilibrium strain which is determined by the knot density and when the strain stop the elastomer return reversibly to their original dimensions. In the case of linear polymers with a coil shape the bonds are physic unions [6, 7].

According to the change of strain rate versus stress the response of the material can be categorized as linear, non-linear, or plastic. When linear response take place the material is categorized as a Newtonian. When the material is considered as Newtonian, the stress is linearly proportional to the strain rate. Then the material exhibits a non-linear response to the strain rate, it is categorized as Non Newtonian material. There is also an interesting case where the viscosity decreases as the shear/strain rate remains constant. This kind of materials are known as thixotropic deformation is observed when the stress is independent of the strain rate [2, 3]. In some cases viscoelastic materials behave as rubbers. In fact, in the case of many polymers specially those with crosslinking, rubber elasticity is observed. In these systems hysteresis, stress relaxation and creep take place.

Figure 2.5 represent in a squematically way different types of responses (σ) to change in strain rate.

The area under the plot Fig. 2.6 between both curves of the circle is a hysteresis loop and represent the amount of energy lost, expressed as heat, in loading and unloading cycle and can be represented by $\oint \sigma \, d\varepsilon$ where σ is stress and ε strain.

Some examples of viscoelastic materials include amorphous polymers, semicrystalline polymers, biopolymers, and metals at very high temperatures. Cracking occurs when the strain is applied quickly and outside of the elastic limit [8].

A viscoelastic material is characterized by at least three phenomena: the presence of hysteresis, which is observed on stress–strain curves, stress relaxation which take place where step constant strain causes decreasing stress and creep occurs where step constant stress causes increasing strain.

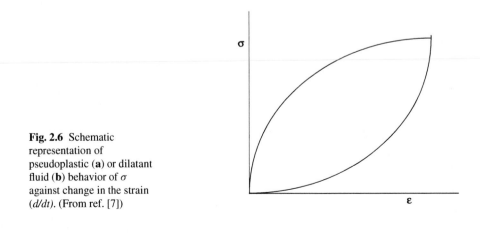

Fig. 2.6 Schematic representation of pseudoplastic (**a**) or dilatant fluid (**b**) behavior of σ against change in the strain (d/dt). (From ref. [7])

Fig. 2.7 Stress–Strain
response for a viscoelastic
material (Newtonian
behavior). (From ref. [7])

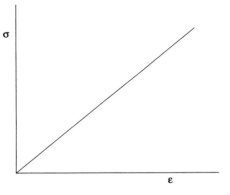

Viscoelastic materials has an elastic component and a viscous component as shown in Fig. 2.7 [7]. The viscosity of a viscoelastic substance gives the substance a strain rate dependent on time [3]. In the case of purely elastic materials no dissipation of energy is observed when a load is applied and then is removed [8]. However, as Meyer et al. [3] pointed out, a viscoelastic substance loses energy when a load is applied and then is removed. In the stress–strain curves, hysteresis is observed within the area of the loop being equal to the energy lost during the loading cycle [3]. Since viscosity is the resistance to thermally activated plastic deformation, a viscous material will lose energy through a loading cycle. Plastic deformation results in lost energy, which is uncharacteristic of a purely elastic material's reaction to a loading cycle [8]. When plastic deformation take place, there is a lost of energy which is uncharacteristic of a purely elastic material's reaction to a loading cycle [3, 8].

Viscoelasticity can be considered as a molecular rearrangement when a stress–strain process take place. In fact when a stress is applied to a polymer the chains suffer a change on their positions which is known as creep.

Creep is the term used to describe the tendency of a material to move or to deform permanently to relieve stresses. Material deformation occurs as a result of long term exposure to levels of stress (physics) that are below the yield strength or ultimate strength of the material. Creep is more severe in materials that are subjected to heat for long periods and near melting point. Creep is often observed in glasses. Creep is a monotonically increasing function of temperature.

The rate of this deformation is a function of the material properties, exposure time, exposure temperature and the applied load (stress). Depending on the magnitude of the applied stress and its duration, the deformation may become so large that a component can no longer perform its function, for example, creep of a turbine blade will cause the blade to contact the casing, resulting in the failure of the blade. Creep is usually of concern to engineers and metallurgists when evaluating components that operate under high stresses or high temperatures. Creep is not necessarily a failure mode, but is instead a deformation mechanism. Moderate creep in concrete is sometimes welcomed because it relieves tensile stresses that otherwise may have led to cracking.

Unlike brittle fracture, creep deformation does not occur suddenly upon the application of stress. Instead, strain accumulates as a result of long-term stress. Creep deformation is "time-dependent" deformation.

The temperature range in which creep deformation may occur differs in various materials. For example, Tungsten requires a temperature in the thousands of degrees before creep deformation can occur while ice formations such as the Antarctic ice cap will creep in freezing temperatures. Generally, the minimum temperature required for creep deformation to occur is 30–40% of the melting point for metals and 40–50% of melting point for ceramics. Virtually any material will creep upon approaching its melting temperature. Since the minimum temperature is relative to melting point, creep can be seen at relatively low temperatures for some materials. Plastics and low-melting-temperature metals, including many solders, creep at room temperature as can be seen marked in old lead hot-water pipes. Planetary ice is often at a high temperature relative to its melting point, and creeps [7].

Creep deformation is important not only in systems where high temperatures are endured such as nuclear power plants, jet engines and heat exchangers, but also in the design of many everyday objects. For example, metal paper clips are stronger than plastic ones because plastics creep at room temperatures. Aging glass windows are often erroneously used as an example of this phenomenon: creep would only occur at temperatures above the glass transition temperature around 900°F/500°C.

An example of an application involving creep deformation is the design of tungsten lightbulb filaments. Sagging of the filament coil between its supports increases with time due to creep deformation caused by the weight of the filament itself. If too much deformation occurs, the adjacent turns of the coil touch one another, causing an electrical short and local overheating, which quickly leads to failure of the filament. The coil geometry and supports are therefore designed to limit the stresses caused by the weight of the filament, and a special tungsten alloy with small amounts of oxygen trapped in the crystallite grain boundaries is used to slow the rate of coble creep.

In steam turbine power plants, steam pipes carry super heated vapor under high temperature (1050°F/565.5°C) and high pressure often at 3500 psiMPa or greater. In a jet engine temperatures may reach to 1000°C, which may initiate creep deformation in a weak zone. For these reasons, it is crucial for public and operational safety to understand creep deformation behavior of engineering materials.

In the initial stage, known as primary creep, the strain rate is relatively high, but slows with increasing strain. The strain rate eventually reaches a minimum and becomes near-constant. This is known as secondary or steady-state creep. This stage is the most understood. The characterized "creep strain rate", typically refers to the rate in this secondary stage. The stress dependence of this rate depends on the creep mechanism. In tertiary creep, the strain-rate exponentially increases with strain [1–9].

2.3.1 The Mechanism of Creep Depends on Temperature and Stress

As Meyers el al. [3] point out, polymers remain as solid material even when these parts of their chains are rearranging in order to accompany the stress, and as this occurs, it creates a back stress in the material. When the back stress is of the same magnitude as the applied stress, the material no longer creeps. When the original stress is taken away, the accumulated back stresses will cause the polymer to return to its original form. The material creeps, which gives the prefix visco, and the material fully recovers, which gives the suffix elasticity.

Viscoelastic materials can follow at least three different behaviors, i.e. linear viscoleasticity, nonlinear viscoelasticity and anelastic behaviour.

Linear viscosity is that when the function is splitted in both creep response and load. All linear viscoelastic models can be represented by a classical Volterra equation connecting stress and strain [1–9]:

$$\varepsilon(t) = \frac{\sigma(t)}{E_{inst,creep}} + \int_0^t K\left(t - t'\right) \sigma\left(t'\right) dt' \tag{2.13}$$

or

$$\sigma(t) = E_{ins,relax}\, \varepsilon(t) + \int_0^t F\left(t - t'\right) \varepsilon\left(t'\right) dt' \tag{2.14}$$

where t is the time, $\sigma(t)$ is the stress, $\varepsilon(t)$ is the strain, $E_{inst,creep}$ and $E_{inst,relax}$ are instantaneous elastic moduli for creep and relaxation, $K(t)$ is the creep and $F(t)$ is the relaxation function [8].

Linear Viscoelasticity is usually applicable only for small deformations [2, 3, 5, 23–25].

The analysis viscoelasticity performed by David Roylance [25] is a nice outline about the mechanical response of polymer materials. This author consider that viscoelastic response is often used as a probe in polymer science, since it is sensitive to the material's chemistry and microstructure [25]. While not all polymers are viscoelastic to any practical extent, even fewer are linearly viscoelastic [24, 25], this theory provide a usable engineering approximation for many applications in polymer and composites engineering. Even in instances requiring more elaborate treatments, the linear viscoelastic theory is a useful starting point.

2.3.2 Molecular Mechanisms

When subjected to an applied stress, polymers may deform by either or both of two fundamentally different atomistic mechanisms. The lengths and angles of the chemical bonds connecting the atoms may distort, moving the atoms to new positions of greater internal energy. This is a small motion and occurs very quickly, requiring only $\sim 10\,[-12]$ seconds [25].

According to Roylance [25] if the polymer has sufficient molecular mobility, larger-scale rearrangements of the atoms may also be possible. For instance, the relatively facile rotation around backbone carbon-carbon single bonds can produce large changes in the conformation of the molecule. Depending on the mobility, a polymer molecule can extend itself in the direction of the applied stress, which decreases its conformational entropy (the molecule is less disordered). Elastomers – rubber- response almost wholly by this entropic mechanism, with little distortion of their covalent bonds or change in their internal energy [25].

The combined first and second laws of thermodynamics state how an increment of mechanical work (fdX = ∂W) done on the system can produce an increase in the internal energy dE or a decrease in the entropy:

$$\partial W = dE - TdS \qquad (2.15)$$

Clearly, the relative importance of the entropic contribution increases with temperature T, and this provide a convenient means of determining experimentally whether the material's stiffnessin in energenic or entropic origin. The retractive force needed to hold a rubber band at fixed elongation will increase with increasing temperature, as the increased thermal agitation will make the internal structure more vigorous in its natural attempts to restore randomness. But the retractive force in a stretched steel specimen- which shows little entropic elasticity-will decrease with temperature, as thermal expansion will act to relieve the internal stress [25].

As Roylance [25] pointed out, in contrast to the instantaneous nature of the energetically controlles elasticity, the conformational or entropic changes are processed whose rates are sensitive to the local molecular mobility. This mobility is influenced by a variety of physical and chemical factors, such as molecular architecture, temperature, or the presence of absorbed fluids which may swell the polymer. Often a simple mental picture of "free volume"- roughly, the space available for molecular segments to act cooperatively so as to carry out the motion or reaction in question- is useful in intuiting these rates.

These rates of conformational change can often be described with reasonable accuracy by Arrhenius-type expressions of the form:

$$v = e^{-\frac{E^{\ddagger}}{RT}} \qquad (2.16)$$

where $E^{\ddagger}$ is the apparent activation energy of the process, R is the gas constant and T the temperature. At temperatures much above the glass transition temperature (Tg), the rates are so fast as to be essentially instantaneous, and the polymer acts in a rubbery manner in which it exhibits large, instantaneous, and fully reversible strain in response to an applied stress.

At temperatures less than Tg, the rates are so low as to be negligible. The chain uncooling process is essentially "frozen out", so the polymer is able to respond only by bond stretching. Therefore at this point the material responds in a "glassy" manner, responding instantaneously and reversibly but being incapable of being strained beyond a few percent before fracturing in a brittle way [25].

Following the Roylance [25] description of the viscoelastic phenomena, in the range near Tg, the material is midway between the glassy and rubbery regimes. Its response is a combination of viscous fluidity and elastic solidity, and this region is termed "Leathery", or, more technically, "viscoelastic". The value of Tg, is an important descriptor of polymer thermomechanical response, and is a fundamental measure of the material's propensity for mobility. Factors that enhance mobility, such as absorbed diluents, expansive stress states, and lack of bulky molecular groups, all tend to produce lower values of Tg. The transparent poly(vinylbutyral) film used in automobile windshielld laminates is an example of material that is used in the viscoelastic regime, as viscoelastic response can be a source of substantial energy dissipation during impact [25].

At temperatures well below Tg, when entropic motions are frozen and only elastic bond deformations are possible, polymers exhibit a relatively high modulus, called the glassy modulus (Eg) which is on the order of 3 Gpa. As the temperature is increased through Tg the stiffness drops dramatically, by perhaps two orders of magnitude, to a value called rubbery modulus Er. In elastomers that have been permanently crosslinked by sulphur vulcanization or other means, the values of Er, is determined primarily by the crosslink density; the kinetics theory of rubber elasticity gives the relation as

$$\sigma = \mathrm{NRT}\left(\lambda - \frac{1}{\lambda^2}\right) \tag{2.17}$$

where σ is the stress, N is the crosslink density (mol/m^3), and $\lambda = \frac{L}{L_0}$ is the extension ratio [25].

Nonlinear viscoelasticity is when the function cannot be splitted. This behaviour is observed when the deformations are large or if the material changes is properties under deformations

Anelastic material is a special case of viscoelasticity in the sense that this kind of materials do not fully recover its original state on the removal of load.

2.3.3 Dynamic Modulus

Viscoelasticity is studied using dynamic mechanical analysis. When we apply a small oscillatory strain and measure the resulting stress. Purely elastic materials have stress and strain in phase, so that the response of one caused by the other is immediate. In purely viscous material the phase delay between stress and strain reach 90 degree phase lag.

Viscoelastic materials exhibit behavior somewhere in the middle of these two types of material, exhibiting some lag in strain [1, 5].

Following the classical treatments of the dynamic modulus G, it can be used to represent the relations between the oscillating stress and strain:

$$G = G' + iG'' \tag{2.18}$$

where $i = \sqrt{-1}$; G' is the *storage modulus* and G″ is the loss modulus:

$$G' = \frac{\sigma_0}{\varepsilon_0} \cos \delta \tag{2.19}$$

$$G'' = \frac{\sigma_0}{\varepsilon_0} \sin \delta \tag{2.20}$$

Where σ_0 and ε_0 are the amplitudes of stress and strain and δ is the phase shift between them [1,5].

Viscoelastic materials, such as amorphous polymers, semicrystalline polymers, and biopolymers, can be modeled in order to determine their stress or strain interactions as well as their temporal dependencies. These models, which include the Maxwell model, the Kelvin-Voigt model, and the Standard Linear Solid Model, are used to predict a material's response under different loading conditions. Viscoelastic behavior is comprised of elastic and viscous components modeled as linear combinations of springs and dashpots, respectively. Each model differs in the arrangement of these elements, and all of these viscoelastic models can be equivalently modeled as electrical circuits. The elastic modulus of a spring is analogous to a circuit's capacitance (in stores energy) and the viscosity of a dashpot to a circuit's resistance (it dissipate energy) [2, 3]. The elastic components, as previously mentioned, can be modeled as springs of elastic constant E, given by equation (2.2) where σ is the stress, E is the elastic modulus of the viscoelastic material and ε is the strain that take place under the given stress. This can be considered similar to Hooke's Law. The viscous components can be modeled as dashpots such that the stress–strain rate relationship can be given as:

$$\sigma = \eta \frac{d\varepsilon}{dt} \tag{2.21}$$

where σ is stress, η is the viscosity of the material and $d\varepsilon/dt$ is the time derivative of strain.

The relationship between stress and strain can be simplified for specific stress rates. For high stress states/short time periods, the time derivative components of the stress–strain relationship dominate. A dashpots resists changes in length, and in a high stress state it can be approximated as a rigid rod. Since a rigid rod cannot be stretched past its original length, no strain is added to the system [23–26].

Conversely, for low stress states/longer time periods, the time derivative components are negligible and the dashpot can be effectively removed from the system – an "open" circuit. As a result, only the spring connected in parallel to the dashpot will contribute to the total strain in the system [23–26].

There are several models to describe the viscoelastic behavior of different materials. Maxwell model, Kelvin-Voigt model, Standard Linear Solid model and Generalized Maxwell models are the most frequently applied.

The Maxwell model can be represented by a purely viscous damper and a purely elastic spring connected in series, as shown in the diagram. The model can be represented by the following equation:

$$\frac{d\varepsilon_{Total}}{dt} = \frac{d\varepsilon_D}{dt} + \frac{d\varepsilon_S}{dt} = \frac{\sigma}{\eta} + \frac{1}{E}\frac{d\sigma}{dt} \tag{2.22}$$

The model represents a liquid (able to have irreversible deformations) with some additional reversible (elastic) deformations. If put under a constant strain, the stresses gradually relax. When a material is put under a constant stress, the strain has two components as per the Maxwell Model. First, an elastic component occurs instantaneously, corresponding to the spring, and relaxes immediately upon release of the stress. The second is a viscous component that grows with time as long as the stress is applied. The Maxwell model predicts that stress decays exponentially with time, which is accurate for most polymers. It is important to note limitations of such a model, as it is unable to predict creep in materials based on a simple dashpot and spring connected in series. The Maxwell model for creep or constant-stress conditions postulates that strain will increase linearly with time. However, polymers for the most part show the strain rate to be decreasing with time [23–26].

The Kelvin-Voigt model, also known as the Voigt model, consists of a Newtonian damper and Hookean elastic spring connected in parallel, as shown in the picture. It is used to explain the stress relaxation behaviors of polymers.

The constitutive relation is expressed as a linear first-order differential equation:

$$\sigma(t) = E_\varepsilon(t) + \eta \frac{d\varepsilon(t)}{dt} \tag{2.23}$$

This model represents a solid undergoing reversible, viscoelastic strain. Upon application of a constant stress, the material deforms at a decreasing rate, asymptotically approaching the steady-state strain. When the stress is released, the material gradually relaxes to its undeformed state. At constant stress (creep), the Model is quite realistic as it predicts strain to tend to σ/E as time continues to infinity. Similar to the Maxwell model, the Kelvin-Voigt Model also has limitations. The model is extremely good with modelling creep in materials, but with regards to relaxation the model is much less accurate.

The Standard Linear Solid Model combines the Maxwell Model and a like Hook spring in parallel. A viscous material is modeled as a spring and a dashpot in series with each other, both of which other, both of which are in parallel with a lone spring. For this model, the governing constitutive relation is:

$$\frac{d\varepsilon}{dt} = \frac{\dfrac{E_2}{\eta}\left(\dfrac{\eta}{E_2}\dfrac{d\sigma}{dt} + \sigma - E_1\varepsilon\right)}{E_1 + E_2} \tag{2.24}$$

Therefore under a constant stress, the modeled material will instantaneously deform to some strain, which is the elastic portion of the strain, and after that it will continue to deform and asynptotically approach a steady-state strain. This last portion is the viscous part of the strain. Although the Standard Linear Solid Model is more accurate than the Maxwell and Kelvin-Voigt models in predicting material responses, mathematically it returns inaccurate results for strain under specific loading conditions and is rather difficult to calculate.

The Generalized Maxwell also known as the Maxwell-Weichert model is the most general form of the models described above. It takes into account that relaxation does not occur at a single time, but at a distribution of times. Due to molecular segments of different lengths with shorter ones contributing less than longer ones, there is a varying time distribution. The Weichert model shows this by having as many spring-dashpot Maxwell elements as are necessary to accurately represent the distribution.

The effect of the temperature on the viscoelastic behavior is an important factor to be taken into account in the analysis of viscoelastic materials when these are submitted to different field forces. The secondary bonds of a polymer constantly break and reform due to thermal motion. Application of a stress favours some conformations over others, so the molecules of the polymer will gradually "flow" into the favoured conformations over time. Because thermal motion is one factor contributing to the deformation of polymers, viscoelastic properties change with increasing or decreasing temperature. In most cases, the creep modulus, defined as the ratio of applied stress to the time-dependent strain, decreases with increasing temperature. Generally an increase in temperature correlates to a logarithmic decrease in the time required to impart equal strain under a constant stress. In other words, it takes less energy to stretch a viscoelastic material an equal distance at a higher temperature than it does at a lower temperature.

When subjected to a step constant stress, viscoelastic materials experience a time-dependent increase in strain. This phenomenon is known as viscoelastic creep.

At a time t_0, a viscoelastic material is loaded with a constant stress that is maintained for a sufficiently long time period. The material responds to the stress with a strain that increases until the material ultimately fails. When the stress is maintained for a shorter time period, the material undergoes an initial strain until a time t_1, after which the strain immediately decreases (discontinuity) then gradually decreases at times $t > t_1$ to a residual strain [2, 23–26].

Viscoelastic creep data can be presented by plotting the creep modulus (constant applied stress divided by total strain at a particular time) as a function of time [23–26]. Below its critical stress, the viscoelastic creep modulus is independent of stress applied. A family of curves describing strain versus time response to various applied stress may be represented by a single viscoelastic creep modulus versus time curve if the applied stresses are below the material's critical stress value.

Viscoelastic creep is important when considering long-term structural design. Given loading and temperature conditions, designers can choose materials that best suit component lifetimes.

Mechanical and viscoelastic behaviour of materials can be determined by different kind of instrumental techniques. Broadband viscoelastic spectroscopy (BVS) and resonant ultrasound spectroscopy (RUS) are more commonly used to test viscoelastic behavior because they can be used above and below room temperatures and are more specific to testing viscoelasticity. These two instruments employ a damping mechanism at various frequencies and time ranges with no appeal to time-temperature superposition. Using BVS and RUS to study the mechanical properties of materials is important to understanding how a material exhibiting viscoelasticity will perform.

2.4 Viscoelastic Properties of Poly(methacrylate)s

Poly(methacrylate)s are a very well known family of vinyl polymers which have the general structures shown in Scheme 2.1:

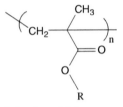

Scheme 2.1 General structure of poly(methacrylate)s

where R can be a broad variety of organic and inorganic substituents. Therefore, this is an interesting family of polymers in order to obtain materials with different characteristics and different behaviors. This is an important aspect from viscoelastic behavior, because it is possible to obtain a great variety of polymers with small differences in their chemical structures and therefore to study the effect of the side chain structure on the viscoelastic behavior of these kind of materials. This is relevant because through the preparation of different families of polymers, the comparison of the viscoelastic responses is a powerfull tool to know the origin of the molecular motions responsibles of the fast relaxations processess that take place when the material is submitted to different field forces. On the other hand, the small differences in the chemical structures allow to know the influence of the fine structure on the response of polymer chains to perturbation fields [27]. By this way the analysis of the viscoelastic responses of the polymeric materials can be performed with some accurate approximation. There are several works dealing with the viscoelastic responses of poly(methacrylate)s containing a great variety of substituents as side chains which are the responsible of the secondary relaxation processes. The effect of the side chain structure on the viscoelastic responses of poly(methacrylate)s have been exhaustively studied and a great amount of information is available about these systems [27–65].

By this way in the case of poly(methacrylate)s it is necessary to split at least two kinds of polymers; i.e. polymers containing aromatic and substituted aromatic groups and those containing aliphatic and substituted aliphatic groups. According to the nature of the chemical structure of the side chain in this family of polymers the viscoelastic response is rather different. For this reason the synthesis, characterization and viscoelastic analysis of poly(methacrylate)s with small differences in the chemical structure in the side chain are a powerfull tool to get information about the origin of the molecular motions responsible of the relaxations processes observed in these systems. Another aspect to be taken into account in the viscoelastic analysis of this kind of materials is that the molecular responses allow to understand the conformational behavior of the polymers and therefore to be able to know which could be their behavior under the application of different force fields under certain conditions. This kind of analysis are very important from technological point of view because according to these results it could be possible to be able to manipulate this materials for specific applications.

There are several works dealing with the dielectric and dynamomechanical behavior of poly(methacrylate)s which can be splitted among those containing aliphatic, saturated cyclic rings and aromatic substituents [27–65].

2.4.1 Poly(methacrylate)s Containing Aliphatic and Substituted Aliphatic Side Chains

The dielectric and viscoelastic properties of poly(methacrylate)s have been extensively studied in the past decades [51–58]. As a general comment polymers containing aliphatic side chains do not present important dielectric and mechanical relaxations. Nevertheless some of these polymers show interesting responses when they are submitted to different field forces. However, it is difficult to find in the literature works in which viscoelastic and dielectric properties of a family of polymers have been studied in a systematic and comparative way. One of the scarce group of poly(methacrylate)s containing substituted aliphatic side chain are poly(2-chloroethyl methacrylate) (P2CEM) and poly(3-chloropropyl methacrylate) (P3CPM) which show important conductive and dipolar components in the dielectric spectrum [40]. In this case, Pelissou et al. [40] have developed a special effort to split conductive and dipolar components in the dielectric spectrum, improving a procedure for this purpose. In this system the whole spectra, ε' and ε'' against temperature is that represented in Fig. 2.8a for P2CEM and P3CPM. The viscoleastic spectra, E' and imaginary one E'' at 1 Hz is shown in Fig. 2.8b. [40]. Three relaxation phenomena are observed from dielectric and viscoelastic spectra. At high temperature an α relaxation associated to the glass transition can be detected, which is partially covered in the dielectric experiments by the increase of dielectric loss ε'' at high temperature attributed to conduction process. β and γ processes are also observed. The position of the γ dielectric relaxation in these systems are in good agreement with those reported by Mikhailov [53]. However the activation energies calculated by Mikhailov [53] are significantly higher. Another aspect to be taken into account in these systems is the analysis of the dielectric behavior at high temperatures for P2CEM and P3CPM which is characterized by a strong increase of dielectric loss ε'' and permittivity ε'.

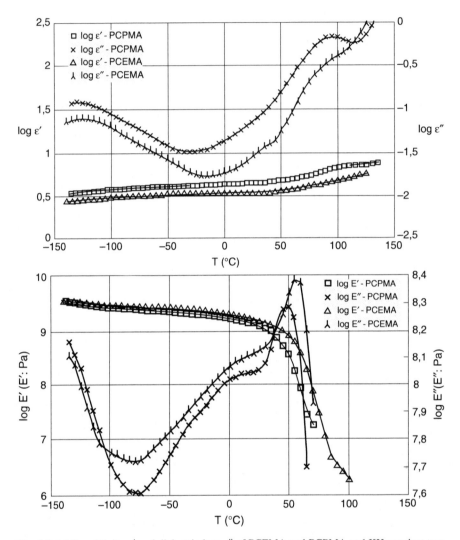

Fig. 2.8 (a) Permittivity ε' and dielectric loss ε'' of PCEMA and PCPMA at 1 KHz against temperature. (b) Storage E' and los modulus E" of PCEMA and PCPMA at 1 Hz against temperature. (From ref. [40])

An interesting method to describe the conductivity contribution can be performed using the modulus M* defined as

$$M^* = \left(\varepsilon^*\right)^{-1} \tag{2.24}$$

This is a procedure which was used successfully by Pathmanathan and Johari and allow to calculate the real an imaginary part of electrical modulus and corresponding spectra as shown in Fig. 2.9

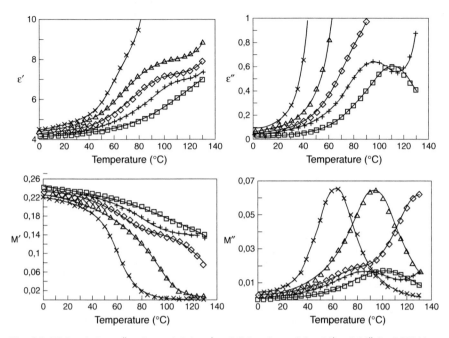

Fig. 2.9 Dielectric loss ε'' and permittivity ε' and dielectric modulus M' and M'' for PCPMA at (X) 1, (Δ) 10, ($\diamond$) 100, (+) 1000 ($\square$) 10000 Hz. (From ref. [40])

In these systems which are very simple from chemical point of view the viscoelastic responses are very complexes but what is the main contribution is the splitting of the dipolar relaxations from those of conductive origin in the dielectric spectra The striking consequence of this transformation is that M"(T) spectra exhibit two families of peacks, attributed to conductive and dipolar relaxations which can be analyzed following different procedures [40]

2.4.2 Poly(methacrylate)s Containing Saturated Cyclic Side Chains

2.4.2.1 Effect of the Spacer Group on Dynamic Mechanical and Dielectric Absorptions

Polymers containing cyclic saturated side chains is a very interesting family of polymers from dielectric and dynamomechanical point of view. These polymers show several relaxation processes due to the conformational versatility and because they are able to present several conformational states [27–33, 66, 67]. The chains have a large number of degrees of freedom which can produce several molecular motions. This structural fact, produces a great variety of transitions and relaxations when the material is affected by mechanical or dielectric force fields. Moreover, the flexibility of the saturated rings also allows to flipping-chair-to-chair –motions of the

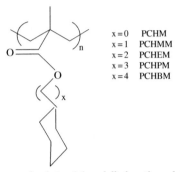

x = 0	PCHM
x = 1	PCHMM
x = 2	PCHEM
x = 3	PCHPM
x = 4	PCHBM

Scheme 2.2 Chemical structures of poly(cyclohexylalkyl methacrylate)s

cyclohexyl group [33]. These motions have been attributed as the responsible of the molecular origin that produce rapid relaxation processes in dynamic mechanical as well as dielectric measurements. From this interpretation, Heijboer [27–33] suggests that the ostensible subglass absorption exhibited by the mechanical spectra of polymers with cyclohexyl groups is produced by flipping motions of the ring. It is very well known that the chair-to-chair motions in the cyclohexyl ring produces a mechanical relaxation at −80°C at I Hz as reported by Heijboer [28]. A mechanism of this type was also suggested to explain the subglass absorption appearing in the dielectric relaxation spectrum of poly(2-chlorocyclohexylacrylate) (P2ClCHA) [66]. However, NMR studies [67] and molecular dynamic calculations [68] performed on poly(2-chlorocyclohexyl isobutyrate), a model compound of P2ClCHA, show that chair-to-chair transition on the cyclohexyl group could not be held the only responsible for the subglass absorption observed in the dielectric relaxation spectra of the polymer at −80°C (1 Hz) [48]. The most simple poly(methacrylate) containing saturated cyclic side chain is poly(cyclohexyl methacrylate) (PCHM) which dielectric and mechanical behavior have been described by Díaz- Calleja et al. [29, 30]. PCHM shows a variety of absorptions due to the versatility of its structural moiety. The effect of the flexible spacer groups on the dynamic mechanical and dielectric behavior of polymers have been taken into account in order to gain confidence about the molecular origin of the relaxations in this kind of materials. By this way it is interesting to analyze the results dealing with polymers containing the cyclohexyl group with aliphatic spacers with different lengths.

(a) Dynamic Mechanical Relaxational Behavior
Figure 2.10 shows the variation of storage and loss modulus for poly(cyclohexyl methacrylate)s where different aliphatic spacer groups are inserted between the cyclohexyl ring and the backbone, i.e. poly(cyclohexyl ethyl, propyl and buty methacrylate)s (PCHEM, PCHPM and PCHBM. In this figure which is a good example of the conformational versatility of the cyclohexyl groups, four relaxation peaks can be observed. At low temperatures there are two relaxations known as δ and γ, respectively. Another two relaxations are observed at high temperatures, known as α and β, the first one correspond to the glass transition temperature. The

Fig. 2.10 Storage modulus
and loss modulus for
PCHEM (●), PCHPM (■) and
PCHBM (▲) at 10 Hz. (From
ref. [33])

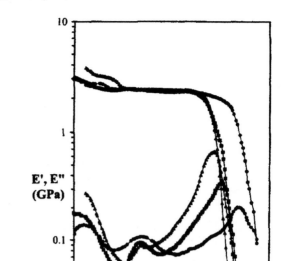

β relaxation is observed as a shoulder of the α relaxation. This behavior preclude
the possibility to perform an exhaustive analysis of the β relaxation [33]. The δ and
γ relaxations are commonly deconvoluted for the Fuoss-Kirkwood [69] empirical
expression:

$$E = \sum_{i=\delta,\gamma} E''''Max_i \sec h\left(m_i \frac{E_a}{R}\left(\frac{1}{T} - \frac{1}{T_{\max}}\right)\right)$$ (2.25)

Where E_a is the activation energy, R is the gas constant, $T_{\max}$ is the temperature
where E'' has a maximum value ($E_{\max}$) and mis a parameter ($0 < m \le 1$) dealing
with the broadness of the peak. On the other hand the strength of the two relaxations
(ΔE) can be calculated using the equation:

$$\Delta E = \frac{2E''\max}{m}$$ (2.26)

In the case of PCHEM and PCHPM the values of m parameter are compiled on
Tables 2.1 and 2.2. These values shows that the γ relaxation is broader than the δ
relaxation. This result is indicative that probably the γ relaxation involve a more
complex molecular motion than δ relaxation.

Table 2.1 Parameters of Huoss-Kirkwood equation for δ and γ relaxations of PCHEM. (From ref. [33])

f (Hz)	$E''_{\max \delta}$ (GPa)	m_δ	$T_{\max \delta}$ (K)	ΔE_δ (GPa)	$E''_{\max \gamma}$ (GPa)	m_γ	$T_{\max \gamma}$ (K)	ΔE_γ (GPa)
30	0.142	0.20	139	1.42	0.080	0.37	221	0.43
10	0.139	0.20	135	1.39	0.083	0.33	213	0.51
3	0.132	0.20	134	1.31	0.081	0.44	198	0.37
1	0.128	0.18	131	1.44	0.082	0.50	192	0.33
0.3	0.127	0.17	128	1.52	0.087	0.48	186	0.37

Table 2.2 Parameters of Fuoss-Kirkwood equation for δ and γ relaxations of PCHPM. (From ref. [33])

f (Hz)	$E''_{\max \delta}$ (GPa)	m_δ	$T_{\max \delta}$ (K)	ΔE_δ (GPa)	$E''_{\max \gamma}$ (GPa)	m_γ	$T_{\max \gamma}$ (K)	ΔE_γ (GPa)
30	0.444	0.29	134	3.03	0.185	0.67	220	0.67
10	0.444	0.30	131	3.00	0.188	0.67	210	0.67
3	0.424	0.30	128	2.80	0.192	0.64	202	0.64
1	0.418	0.31	126	2.72	0.199	0.57	196	0.57
0.3	0.401	0.29	123	2.75	0.209	0.52	198	0.52

Table 2.3 Activation energy and tan $\delta_{\max}$ (10 Hz) for δ and γ relaxations. (From ref. [33])

Polymer	$E_{a\delta}$ (kcal mol^{-1})	$E_{a\gamma}$ (kcal mol^{-1})	tan $\delta_{\max \delta}$	tan $\delta_{\max \gamma}$
PCHM[1]		11.3		0.081
PCHMM[3]		12.5		0.051
PCHEM	9.06	11.19	0.05	0.045
PCHPM	6.46	10.93	0.063	0.040
PCHBM		9.61		0.037

The activation energies for δ and γ relaxations for PCHEM, PCHPM and PCHBM obtained from Arrhenius plots are summarized on Table 2.3. The results compiled in this Table are very interesting in the sense that a decreasing of the activation energy with increasing of the length of the spacer group is observed.

Therefore, the effect of the length of the spacer group play an important role in the viscoleastic behavior of this family of polymers.

In this family of polymers, with saturated cyclic side chain the effect of the side chain structure and also the effect of the spacer group on the viscoelastic behaviour is clearly illustrated in Fig. 2.11a and b in which it is possible to observe the variation of tan δ for five members of the series with temperature for δ and γ relaxations.

Clearly the maximum of tan δ value for the γ relaxation decreases as the length of the spacer group increases. Instead, the maximum of the loss tangent of the δ relaxataion increases as the length of the spacer group increases. δ relaxation is not observed in the polymer without spacer group, i.e. poly(cyclohexyl methacrylate) [28, 33]. This relaxation is attributed to another kind of motion of the cyclohexyl ring like small rotations of the group as a whole around the bond containing the connecting link. If the spacer is absent as in the case poly(cyclohexyl methacrylate), those motions would be restrained. Therefore the spacer group participate in both

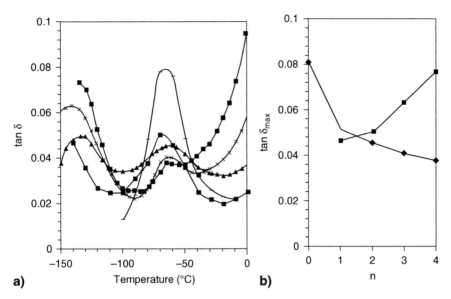

Fig. 2.11 (a) Temperature dependence of the tan δ at 10 Hz for (−) PCHM (From ref. [28]), (■) (From ref. [29]), (▲) PCHEM, (X) PCHPM and (○) PCHBM. (b) Variation of the intensity of the maximum of tan δ With the number of carbon atoms of the spacer group for the γ (♦) and δ (■) relaxations. (From ref. [33])

the δ and γ relaxations. These results are interesting demonstrations that the effect and nature of the side chain of the polymers play an important role on the dynamic mechanical and dielectric behaviour of the material [33].

(b) Dielectric relaxational behavior

The dielectric analysis of these systems show, that only one peak can be observed corresponding to the dynamic glass transition. The sub-glass relaxations are very small.

The α relaxation without the conductive contribution can be analyzed using the Havriliak-Negami (HN) equation [70]:

$$\varepsilon^* = \varepsilon_\infty + \frac{\varepsilon_0 - \varepsilon_\infty}{\left(1 + \left(j\omega\tau\right)^{\bar{\alpha}}\right)^{\bar{\beta}}} \tag{2.27}$$

where ε_0 and ε_∞ are the relaxed and unrelaxed permittivity, respectively, and $\bar{\alpha}$ and $\bar{\beta}$ are two parameters related to the shape and skewness of the Cole-Cole plot. For this system the HN parameters are summarized in Table 2.4.

Figure 2.12 represent the plot of $\log \varepsilon''$ against $\log \omega$ where ε'' and ω are the conductive contribution and ω is the angular frequency.

Table 2.4 Parameters of Havriliak-Negami equation (2.8) for α relaxation at indicated temperature. (From ref. [33])

Polymer	T (°C)	ϵ_∞	ϵ_0	$\bar{\alpha}$	$\bar{\beta}$	τ (s)
PCHEM	90	3.02	4.59	0.42	0.70	0.18
PCHEM	95	3.00	4.51	0.44	0.65	0.058
PCHPM	45	2.13	3.46	0.36	0.78	0.040
PCHBM	45	2.14	3.21	0.60	0.64	2.32×10^{-3}
PCHBM	50	2.16	3.17	0.69	0.60	1.05×10^{-3}

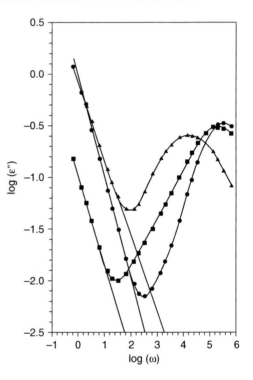

Fig. 2.12 Frequency dependence of the dielectric loss ε'' for (▲) PCHEM at 125°C, (■) PCHPM at 100°C and (●) PCHBM at 90°C. The *solid line* indicate the conductive contribution with different slopes. (From ref. [33])

Here a different tendency for the slope in the conductivity contribution can be observed.

The activation energy of the conductivity in theses systems is calculated using an Arrhenius plot of log σ, the conductivity against T^{-1} shown in Fig. 2.13.

The values of the activation anergy in this case are 24.6, 18.1 and 25.8 kcal mol^{-1} for PCHEM, PCHPM and PCHBM respectively.

(c) Molecular dynamic simulation.

Another way to get information about the relaxational behavior of these materials can be performed by dynamic mechanical calculations. In order to get information about the origin of the secondary γ relaxation, molecular dynamic (MD) calculations over the repeating unit were performed. By this way considering the axial and equatorial equilibrium on the cyclohexyl group and the interconversion of these two

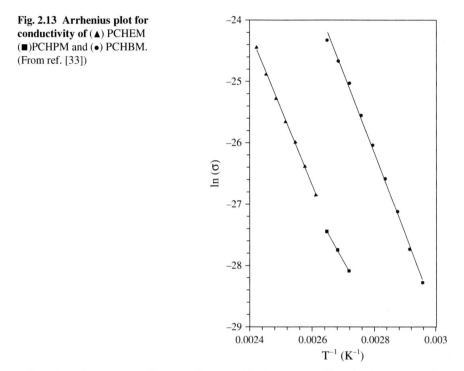

Fig. 2.13 Arrhenius plot for conductivity of (▲) PCHEM (■)PCHPM and (●) PCHBM. (From ref. [33])

orientations is a matter of current interest. For instance at 5 ns the interconversion between axial and equatorial orientation is not observed in for the MD trajectories. After the stabilization of the system it is possible to calculate the ratio of the fractions of axial and equatorial conformations (f_{eq}/f_{ax}) by using the Boltzman equation:

$$\Delta G = -\mathrm{RTLn}\frac{f_{eq}}{f_{ax}} \tag{2.28}$$

The $\dfrac{\Delta G}{R}$ value between these two conformations can be obtained by plotting $Ln\dfrac{f_{ax}}{f_{eq}}$ against T^{-1}.

After calculation using molecular dynamic simulation at several temperatures it is possible to know the effect of the length of the spacer group on the conformational and viscoelastic behavior. Figure 2.14 shows the variation of $\Delta G/R$ for the conformational chair-to-chair flipping against the number of carbon atoms calculated for the γ relaxation. Comparison of this plot with the experimental variation of tanδ vs n (the number of carbon atoms of the spacer group) one Fig. 2.14a, is in good agreement. This comparison is achieved through the analysis of the variation of the intensity of the maximum of tan δ for the γ relaxation (taken from Fig. 2.14b) against $\dfrac{\Delta G}{R}$. A linear correlation between values of $\dfrac{\Delta G}{R}$ and the experimental intensity of tanδ can be obtained, with an straight line shape with a correlation of

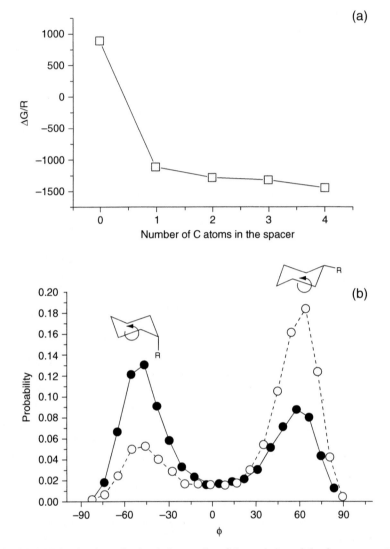

Fig. 2.14 (a) Molecular dynamic simulation results of the variation of the free energy change ($\Delta G/R$) of the chair-to-chair conformational change against of the number of carbon atoms of the spacer group. (b) Probability distribution of the torsional angle ϕ obtained at 1000 K for PCHMA (●) and PCHBM (○). (From ref. [33])

0.982. Therefore, the free energy change against the number of carbon atoms in the side chain is in excellent agreement with the experimental data which means that the number of carbon atoms of the spacer group has an influence on the γ relaxation associated with the chair-to-chair conformational change in the cyclohexyl group [33].

Figure 2.14b shows the results of the probability distribution for the torsional angle ϕ (see scheme in Fig. 2.14) obtained at 1000 K for PCHMA and PCHBMA in order to stress the differences in the conformational behavior of the polymers due to the presence of the spacer group. The areas under the peaks of each curve represent the relative incidence of the two conformations allowed to the molecule and thus permit the evaluation of the fractions f_{ax} and f_{eq}. According to these results, when the side group of the polymer chain does not have a spacer group, i.e., the cyclohexyl group is directly joined to the ester group, axial conformations are preferred, but when the spacer group exists, equatorial conformations are preferred. This molecular dynamic approach is useful to explain in part the experimental differences observed in $\tan\delta_{max}$ for the γ relaxation in poly(cyclohexyl methacrylate)s [33].

These results have been shown that the presence of four distinct mechanical relaxations for these three polymers analyzed, namely δ, γ, β and α in the temperature range from -150 to $100°C$. The behavior of the α relaxation is similar to that reported for polymers with analogous structures. [33]. The α and γ relaxations of PCHEM and PCHPM have been characterized by Fuoss-Kirkwood equations. The maximum of the loss factor in δ relaxation is lower and the relaxation is broader than the γ relaxation. Therefore, the intensity of the δ is larger than the γ relaxation. Nevertheless, the height of the loss factor depends on the length of the spacer group what would indicate that the spoacer group takes part on both the δ and γ relaxations [33], (see Fig. 2.14b) obtained at 100 K for PCHMA and PCHBM in order to stress the differences in the conformational behavior of the polymers due to the presence of a spacer group. The areas under the peaks of each curve present the relative incidence of the two conformations allowed to the molecule and thus permit the evaluation of the fractions f_{ax} and f_{eq}. According to these results, when the side chain does not have a spacer group, i.e., the cyclohexyl group is directly joined to the ester group, axial conformational are preferred, but when the spacer group exists, equatorial conformations are preferred. This molecular dynamic approach seems to be useful to explain in part the experimental differences observed in $\tan\delta$ for the γ relaxation in poly(cyclohexylalkyl methacrylate)s [33, 71].

In this system the α relaxation can be analyzed by the symmetric equation of Fuoss-Kikwood and a new model which is similar to Havriliak- Negami equation used in the analysis of dielectric spectroscopy. According to the Tg values calculated for these systems, the free volume can be appropriately described by the free volume theory. The analysis of these families of poly(methacrylate)s allow to understand in a good way the effect of the structure and nature of the side chain on the viscoleastic behavior of polymers [33].

2.4.2.2 Effect of Bulky Substituents on the Dynamic Mechanical and Dielectric Absorptions

The effect of the side chain structure on the solution behavior as well as in the solid state of vinyl polymers has been studied in the past for a number of poly(methacrylate)s [72, 73, 76]. The conformational study of polymers containing aromatic, aliphatic and saturated cyclic side chains has demonstrated that the nature, volume and chemical structure of the substituents notably influence the conformation and rigidity of the polymer chain [32, 77, 78]. The rigidity of the chain in these kind of polymers depends, in a significant way, on the spatial volume of the side groups and on their specific interactions [74, 75, 79, 80]. Polymers containing bulky side chains show high characteristic ratios C_∞, and rigidity coefficient $\left(r_0^2/r_{0f}^2\right)^{1/2}$, due to the steric hindrance to rotation of the main chain [32, 72, 81]. Owing to the bulky side groups, these polymers also present high Tg values [77, 81]. However, in spite of this characteristic, the mobility of the side chain is large enough to produce dielectric as well as mechanical activity [27, 29, 30, 32, 82, 83]. On the contrary polymers containing cyclohexyl groups are able to show several conformational states [27, 29, 30, 32, 82, 83]. These chains have a large number of degree of freedom, that produce several molecular motions. This structural fact produces a great variety of transitions and relaxations when the material is affected by mechanical and dielectric force fields [32].

The flexibility of the saturated ring also allows flipping (chair-to-chair) motions of the cyclohexyl group which has been attributed as molecular origin that produce rapid relaxation processes in dynamic mechanical and dielectric measurements [32]. It is well known that the chair-to-chair motions in the cyclohexyl ring produce a mechanical relaxations at about $-80°C$ at I Hz, as reported by Heijboer [28]. In the case of poly(cyclohexylmethyl methacrylate) and poly(cyclohexylethylmetacrylate) (PCHEM) the dielectric and mechanical behavior presents a variety of absorptions due to the versatility of their molecular moiety as it was mentioned above [30, 82–85]. The question is open in the sense that if the substitution of bulky groups in the saturated rings inhibit or not this relaxational behavior in this framework, polymers with ter-butyl groups as substituents of hydrogens in the cyclohexyl ring are good candidates to analyze this effect. Two polymeric systems that represents this analysis are poly(2-tert-butylcyclohexyl methacrylate) P2tBCHM and poly

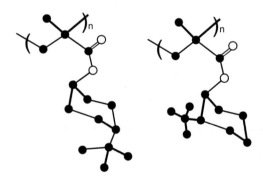

P4tBCHM   P2tBCHM

Scheme 2.3 Chemical structures of P4tBCHM and P2tBCHM. (From ref. [32])

(4-ter-butylcyclohexyl methacrylate) P4tBCHM. Therefore the comparison of the dielectric and dynamic mechanical behavior of poly(2-tert-butyl cycloheyl methacrylate) (P2tBCHM) and poly(4-tert-butylcyclohexylmethacrylate) (P4tBCHM) See (Scheme 2.3) [32]

(a) Dynamic mechanical relaxational behavior

Figures 2.15 and 2.16 show the storage and loss moduli for poly(2-tert-butylcyclohexyl methacrylate) (P2tBCHM) and poly(4-tert butylcycloheyl methacrylate) P4tBCHM) [32].

In these figures, it can be observed that the mechanical loss for P2TBCHM is more complex than that for P4tBCHM, which as can be seen in Figs. 2.15 and 2.16 is relatively featureless except in the zone corresponding to the glass transition temperature (Tg).

Both polymers show a strong relaxation at about 120°C and 100°C for as can be seen in Figs. 2.15 and 2.16, for (P2tBCHM) and, (P4tBCHM) as Díaz Calleja et al. [32] have reported. Moreover P2tBCHM show a complex secondary relaxation at about −80° and a remainder of the mechanical activity at about −20°C and 30°C respectively [32] poly(4-tert butylcyclohexyl methacrylate) (P4tBCHM) respectively.

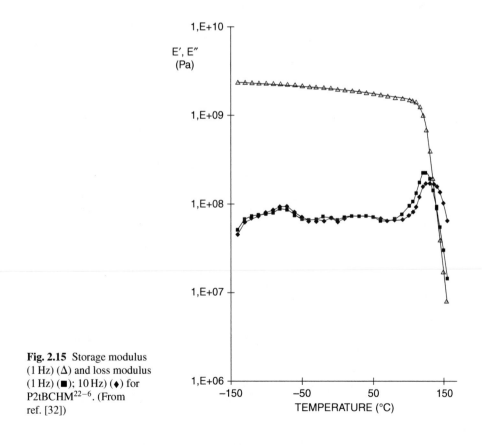

Fig. 2.15 Storage modulus (1 Hz) (Δ) and loss modulus (1 Hz) (■); 10 Hz) (♦) for P2tBCHM[22-6]. (From ref. [32])

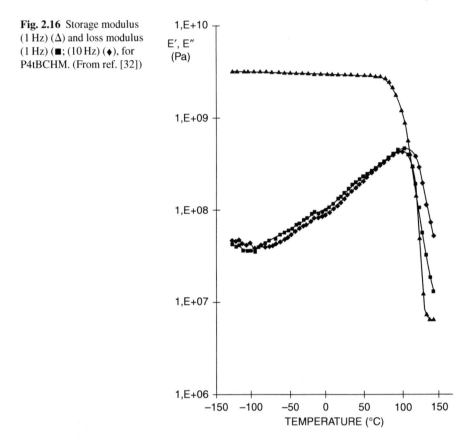

Fig. 2.16 Storage modulus (1 Hz) (Δ) and loss modulus (1 Hz) (■; (10 Hz) (♦), for P4tBCHM. (From ref. [32])

Moreover P2tBCHM show a complex secondary relaxation at about −80°C and a remainder of the mechanical activity vat about −20°C and 30°C respectively. The first of this remaining peaks has a counterpart in P4tBCHM. It is noteworthy that the α relaxation associated with the dynamic glass transition temperature, in both polymers seems to have structure, what is an important aspect to be taken into account in the analysis of the dynamic mechanical responses of these systems. In fact, in both polymers the α relaxation is splitted in with increasing temperature and frequency. At low temperatures and low frequencies in the α zone, only a single peak is apparent [32]. However, with increasing frequency and temperature a new peak appear that becomes dominant at higher temperatures.

(b) Dielectric relaxational behavior

Dielectric permittivity and loss for both polymers under study can be observed on Figs. 2.17 and 2.18. In both figures a prominent peak corresponding to the dynamic glass transition temperature can be observed, which at low frequencies is overlapped by conductivity effects. Moreover, in both polymers a broad secondary peak is observed at about −50°C. This peak is more prominent in P2tBCHM which is in good

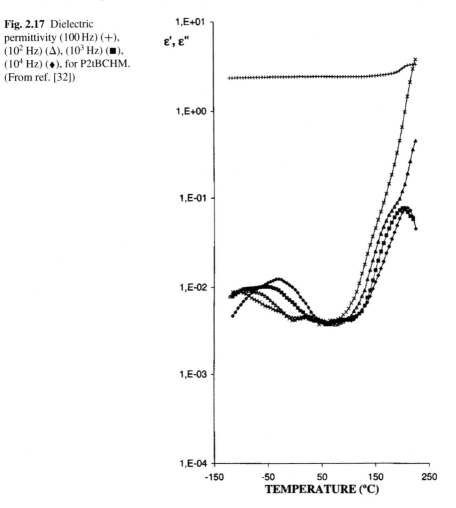

Fig. 2.17 Dielectric permittivity (100 Hz) (+), (10^2 Hz) (Δ), (10^3 Hz) ($\blacksquare$), (10^4 Hz) ($\blacklozenge$), for P2tBCHM. (From ref. [32])

agreement with mechanical measurements. It is possible to observe a remaining dielectric activity in both polymers in the temperature range between 50°C and 100°C. This remaining activity is enhanced by the conductivity in the case of P4tBCHM.

Figure 2.19 show an Arrhenius plot for the dielectric β relaxation for P2tBCHM and P4tBCHM from which the activation energies (E_a) are obtained from the slope of the straight lines. The values obtained are 38.6 ± 0.5 kJ mol^{-1} for the mechanical sub-T_g relaxation of P2tCHM and 46 ± 0.5 kJ mol^{-1} and 39.30.5 kJ mol^{-1} for the dielectric sub-T_g relaxations for P2tBCHM and P4tBCHM respectively.

In the α zone, the conductivity contribution can be separated using a hopping model [86] according to equation:

$$\varepsilon'' = \frac{\sigma}{\varepsilon_0 \omega^s} \quad 0.5 \leq s \leq 1 \qquad (2.29)$$

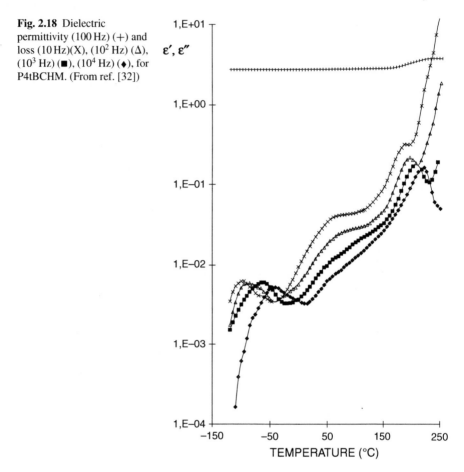

Fig. 2.18 Dielectric permittivity (100 Hz) (+) and loss (10 Hz)(X), (10^2 Hz) (Δ), (10^3 Hz) (■), (10^4 Hz) (♦), for P4tBCHM. (From ref. [32])

where ω is the angular frequency, ε_0 the permittivity of the vacuum and σ the conductivity. The values obtained for each polymer are compiled in Table 2.5 where the bimodal structure of the conductivity can be observed.

The activation energy of the conductivity can be obtained from a plot of Ln σ against T^{-1} as shown in Fig. 2.20, where the presence of two zones presumably associated to the different types of charge carriers are observed. The values of the activation energy are 57 and 162 kJ mol^{-1} for P2tBCHM and 53 and 178 kJ mol^{-1} for P4tBCHM for low and high temperature zones, respectively.

Figures 2.21 and 2.22 are examples of the obtention of a clean α peak after subtracting the conductivity. Afterwards it is possible to fit an empirical Havriliak-Negami equation (87) to the experimental data following the usual procedure.

$$\varepsilon^* = \varepsilon_\infty + \frac{\varepsilon_0 - \varepsilon_\infty}{(1 + (j\omega\tau)^\alpha)^\beta} \qquad \alpha \leq 1, \quad \alpha\beta \leq 1 \qquad (2.30)$$

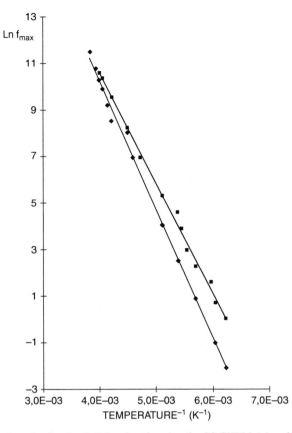

Fig. 2.19 Arrhenius plot for the β dielectric relaxation for P2tCHBM (♦) and P4tBCHM (■). (From ref. [32])

The parameters for P4tBCHM are summarized in Table 2.6

The temperature dependence of the α relaxation in the frequency domain can be conveniently analyzed by means of the Vogel-Fulcher-Tamman-Hesse (VFTH) equation [88–90] which was empirically formulated as:

Table 2.5 s parameter of equation (2.29) to P2tBCHM and P4tBCHM. (From ref. [32])

P2tBCHM			P4tBCHM		
T (°C)	s	σ_0 ($\Omega^{-1}\,m^{-1}$)	T (°C)	s	σ_0 ($\Omega^{-1}\,m^{-1}$)
220	0.9314	1.2819×10^{-9}	220	0.8618	5.2353×10^{-10}
200	0.827	1.8224×10^{-10}	200	0.8909	8.6704×10^{-11}
180	0.6819	2.879×10^{-11}	180	0.6984	2.3288×10^{-11}
160	0.6482	7.4858×10^{-12}	160	0.6504	1.3581×10^{-11}
140	0.6599	3.1589×10^{-12}	140	0.6616	6.6656×10^{-12}

Fig. 2.20 Arrhenius plot for
the conductivity of
P2tBCHM (♦) and P4CHBM
(■) showing the bimodal
structure in each case. (From
ref. [32])

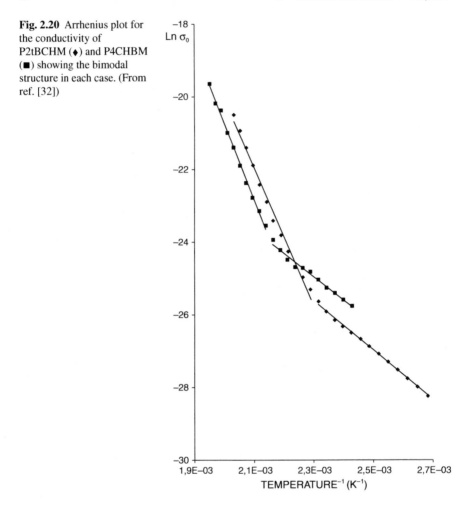

$$\ln f_{\max} = A' - \frac{m'}{T - T_\infty} \tag{2.31}$$

where T_∞ is an empirical parameter related to the Kauzman temperature or the
temperature at which the conformational entropy is zero. According to the best fit
of the dielectric experimental results to equation (2.31) the values for T_∞ are 381
and 373 K for P2tBCHM and P4tBCHM. The values of m' amounts to 2197 2186
for P2tBCHM and P4tBCHM respectively obtained from Fig. 2.23.

Poly(methacrylate)s containing saturated cyclic rings as side chains are a family
of polymers that gives a great amount of information about the relaxation processes
that take place when these polymers are submitted to different dielectric and mechanical force field.

Despite of the rigidity of both polymers under study and the high values observed for the glass transition temperature, significant dielectric subglass activity

Fig. 2.21 Loss permittivity in the α relaxation zone, after subtracting conductivity at different frequencies: (10 Hz) (X); (10^2 Hz) (Δ); (10^3 Hz) ($\blacksquare$); (10^4 Hz) ($\blacklozenge$); for P2tBCHM. (From ref. [32])

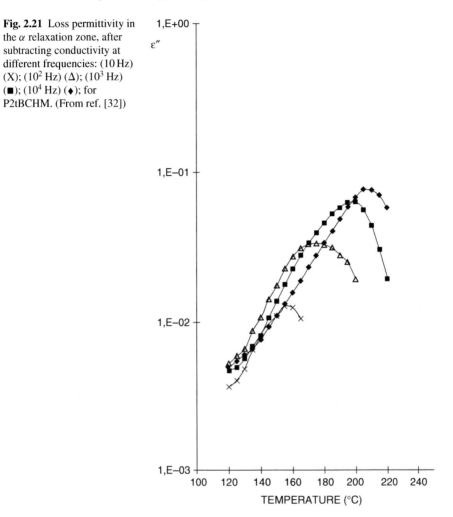

is present in both polymers. In fact, the *tert*-butyl butyl group is essentially located in the equatorial position inhibiting the chair-to-chair motions but nor wholly suppressing. This activity is more important in the case of P2tBCHM. The bond that links a substituent in the 4-position to the ring retains its direction in space, shifting parallel to itself during the transition, as pointed out by Heijboer [28]. However, a substituent in the 2-position is turned through an angle of 109° [28]. As a result, the dielectric effect of the remaining chair-to-chair transition should be much larger for a substituent in 2-position than for one in the 4-position. On the contrary, the mechanical sub-Tg activity is lower than the dielectric one. In the case of P4tBCHM, there are not clearly defined sub-Tg peaks. This means that these two polymers are more easily activated by electric force fields than by mechanical

Fig. 2.22 Loss permittivity in the α relaxation zone, after subtracting conductivity at different frequencies: (10 Hz) (X); (10^2 Hz) (Δ); (10^3 Hz) ($\blacksquare$); (10^4 Hz) ($\blacklozenge$); for P4tBCHM. (From ref. [32])

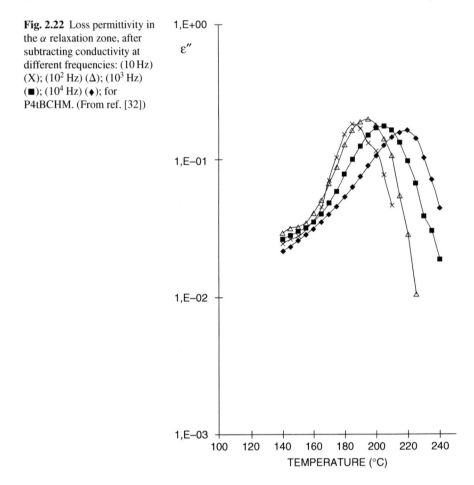

ones. Owing to the fact that the only relevant dipolar group is ester in both polymers, it can be concluded that in the molecular origin of the dielectric subglass relaxation, some contribution of this group is the responsible of this relaxational behavior. This contribution is more important in P2tBCHM than in P4tBCHM for the reason given above. The mechanical subglass activity is also in good agreement with the fact that higher mobility is found in the polymer with the substitution in the 2- position.

Table 2.6 Parameters of de Havriliak-Negami equation for P4tBCHM. (From ref. [32])

$T(°C)$	ε_∞	Δ_ε	α	β	τ_0
195	2.76	1.03	0.55	0.61	0.0032
190	2.77	1.07	0.53	0.60	0.0090

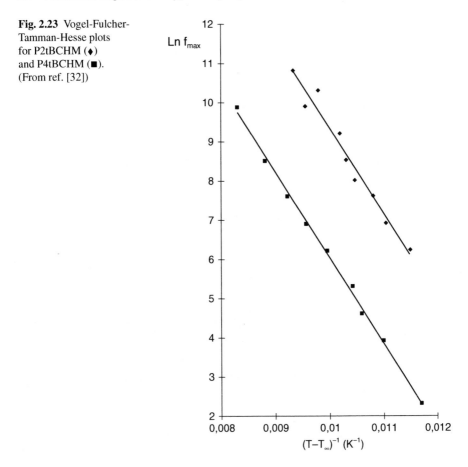

Fig. 2.23 Vogel-Fulcher-Tamman-Hesse plots for P2tBCHM (♦) and P4tBCHM (■). (From ref. [32])

2.4.2.3 Effect of the Number of Members in the Saturated Rings

Poly(cycloheptyl methacrylate)s, Poly(cyclooctyl methacrylate)

(a) Dynamic mechanical relaxational behavior

Heijboer [28] has reported the dynamic mechanical properties of poly(methacrylate)s with different size of the saturated ring as side chain. The γ relaxation in these polymers is attributed to a conformational transition in the saturated ring. In the case of poly(cyclohexyl methacrylate), the transition is between the two chair conformations in the cyclohexyl ring. However, this type of internal motion in hindered by rather high intramolecular barriers, which can reach about 11 kcal mol^{-1}.

Mechanical and dielectric behavior of poly(methacrylate)s with cyclohexyl groups in the side chain have been reported as it was described above and the viscoelastic information obtained from these polymeric systems is very broad and give confidence about the molecular origin of the fast relaxation processes that take place in these systems. [30,32,36,48,82]. As it was described above these polymers show a great variety of absorptions due to the versatility of their structural moieties.

Polymers with larger numbers of carbon atoms should present more relaxational activity due to the increasing versatility of the ring [36]. It has been reported that polymers containing rings with an odd number of carbon atoms show higher relaxational activity at higher temperatures than the corresponding polymers with even-membered rings [57].

The viscoelastic analysis of poly(cycloheptyl methacrylate) (PCHpM), poly (cycloheptylmethyl methacrylate) (PCHpMM) and poly(cyclooctylmethacrylate) (PCOcM) (see Scheme 2.4) is a good example of the relaxational behavior of polymers containing saturated rings in the side chain.

Figures 2.24, 2.25 and 2.26 show the storage and loss tensile moduli for poly (cycloheptyl methacrylate) (PCHpM), poly(cycloheptylmethyl mathacrylate) (PCHpMM) and poly(cyclooctylmethacrylate) (PCOcM), respectively.

The relaxation process associated with the dynamic glass transition, the α relaxation, and the β relaxation, as a shoulder of the α relaxation can be observed in all these figures. At low temperatures another relaxation labeled as γ relaxation can be observed. In the case of PCHpM, the maximum of the γ relaxation is well away from the temperature range. Heijboer and Pineri [36, 57] have reported that the maximum for this polymer is at about 100 K for 1 Hz. In the case of PCHpMM and PCOcM, the γ relaxation can be observed which may be analyzed by using the Fuoss-Kirkwood (F-K) equation:

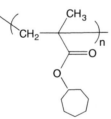

Poly(cycloheptyl methacrylate)
PCHpM

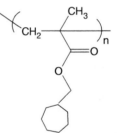

Poly(cycloheptyl methymeyhacrylate)
PCHpMM

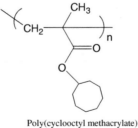

Poly(cyclooctyl methacrylate)
PCOcM

Scheme 2.4 Chemical structures of PCHpM, PCHMM and PCOcM

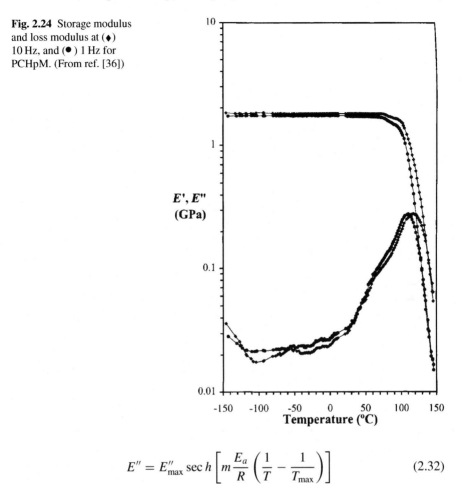

Fig. 2.24 Storage modulus and loss modulus at (♦) 10 Hz, and (●) 1 Hz for PCHpM. (From ref. [36])

$$E'' = E''_{max} \sec h \left[m \frac{E_a}{R} \left(\frac{1}{T} - \frac{1}{T_{max}} \right) \right] \tag{2.32}$$

where E_a is the activation energy, R the gas constant, T_{max} is the temperature at which ε'' reaches to a maximum value (ε''_{max}) and m is a parameter $(0 < m \leq 1)$ which is related with the breadth of the peak. The strength of the relaxation (ΔE) can be calculated by using the equation:

$$\Delta E = \frac{2E''_{max}}{m} \tag{2.33}$$

the parameters obtained at different frequencies are summarized in Tables 2.7 and 2.8.

The activation energy obtained from an Arrhenius plot are 9.0 kcal mol^{-1} for PCHpMM and 10.4 kcal mol^{-1} for PCOcM [36]. Heijboer [28, 57] reported the activation energy for poly(cyclopentyl methacrylate) (PCPM), poly(cyclohexyl methacrylate) (PCHM), PCHpM and PCOcM as 3.1, 11.6, 6.2 and 10.6 kcal mol^{-1} respectively. From the earlier results it can be seen that polymers with odd-membered

Fig. 2.25 Storage modulus at
(♦) 10 Hz and (●) 1 Hz for
PCHpMM. (From ref. [36])

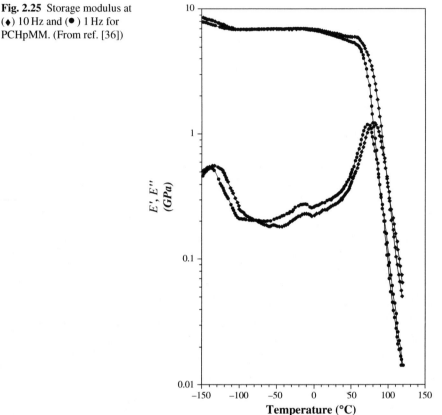

rings have a lower activation energy than the even-membered rings. However, an opposite tendency is observed in the barrier opposing the motions in the sense that the activation energy for polymers with odd membered ring increases with the complexity of the ring while for polymers with even-membered rings decreases.

In the case of the seven-membered ring polymer (PCHpM), the insertion of a methyl group as spacer increases the activation energy and give rise to the appearance of a new peak in the experimental range of measurements which is located between the seven and eight-membered ring poly(methacrylate)s. This means that the insertion of a methyl spacer group have the opposite effect in this case than for the six-membered ring polymers. In the same way this insertion tends to equalize the activation energies of the PCHpMM and PCOcM what is a significative result [36].

The dependence of the α relaxation in the frequency domain can be analyzed using the VFTH theory [88–90] as in the case of poly(t-butycyclohexyl methacrylate)s described above, using equation (2.31). The values obtained for T_∞ are 280, 232 and 253 K for PCHpM, PCHpMM and PCOcM respectively. It is possible to calculate

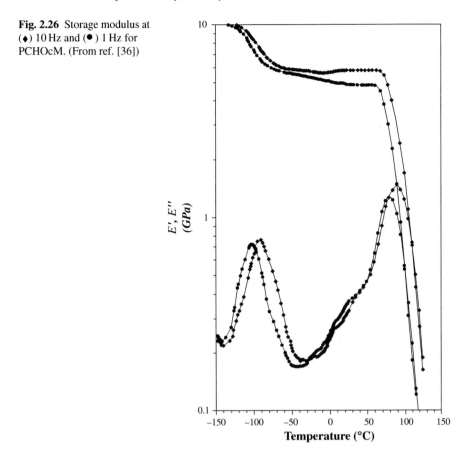

Fig. 2.26 Storage modulus at (♦) 10 Hz and (●) 1 Hz for PCHOcM. (From ref. [36])

the free volume at the gas transition temperature (Tg) according to the following equation:

$$\frac{\phi}{B} = \frac{T - T_\infty}{m'} \qquad (2.34)$$

where the relationship $\frac{\phi}{B}$ is the free volume at T and the values of m' and T_∞ are the parameters of the VFTH equation. The relative free volumes at Tg are 3.2, 3.2 and 3.3 for PCHpM, PCHpMM and PCOcM respectively, which are in agreement with the free volume theory [36].

Table 2.7 Parameters of Fuoss-Kirkwood for γ relaxation of PCHpMM. (From ref. [36])

f (HZ)	E''_{max} (GPa)	m	T_{max} (K)	$\Delta E (GPa)$
30	0.584	0.22	144	4.05
10	0.577	0.22	140	3.96
3	0.550	0.21	136	4.28
1	0.545	0.21	132	4.32

Table 2.8 Parameters of Fuoss-Kirkwood for γ relaxation of PCOcM. (From ref. [36])

f (HZ)	$E''_{max}(GPa)$	m	$T_{max}(K)$	$\Delta E(GPa)$
30	0.703	0.35	188	4.05
10	0.743	0.38	182	3.96
3	0.691	0.32	174	4.28
1	0.708	0.33	169	4.32
0.3	0.741	0.33	163	4.50

(b) Dielectric relaxational behavior

Figures 2.27, 2.28 and 2.29 represents the permittivity and dielectric loss for PCHpM, PCHpMM and PCOcM respectively. In this case, only PCOcM a peak

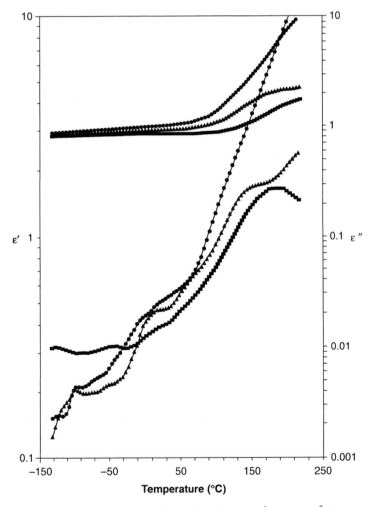

Fig. 2.27 Dielectric permittivity and loss for PCHpM at (■) 10^4 Hz, (▲) 10^2 Hz, and (•) 1 Hz. (From ref. [36])

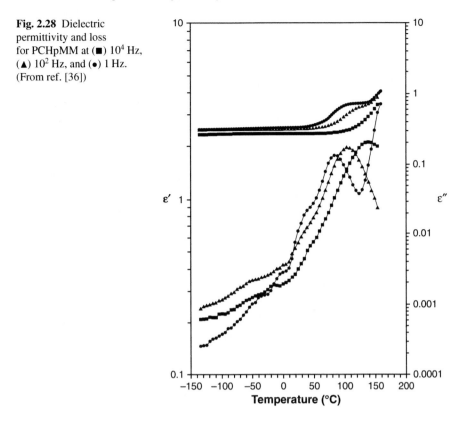

Fig. 2.28 Dielectric permittivity and loss for PCHpMM at (■) 10^4 Hz, (▲) 10^2 Hz, and (●) 1 Hz. (From ref. [36])

corresponding to the sub-glass relaxation is observed. Figure 2.30 show the Arrhenius plot of γ, β and α relaxations for PCOcM. The activation energy for the γ and β relaxations are 10.3 and 29.9 kcal mol^{-1}.

As previously was analyzed, the γ relaxation can be analyzed by the symmetric Fuoss-Kikwood equation [91]. It is interesting to note that below their glass transition PCHpM and CHPMM do not show significant dielectric activity in the range of temperatures described [36].

In all the polymers under study, dipolar α relaxation appear to be overlapped by spurious conductive effects. This is a common feature in many polymers at low frequencies and temperatures above the glass transition [36]. The conductive and blocking electrode contributions to the loss can be conveniently analyzed by several theoretical procedures [92,93]. In these models two different processes for the space charge relaxation are considered: one is due to the conduction of the free species remaining in the polymer and the second is due to the blocking electrodes. The first, the conductive phenomenon presumes the existence in the polymer of a number of mobile charges that, in the absence of an electric field are in equilibrium. When a field is applied, charges are then subjected to the combined influence of the field and the thermal diffusion. The former tend to accumulate charges near the electrodes (blocking electrodes), while the latter tends to oppose this charge accumulation.

Fig. 2.29 Dielectric
permittivity and loss
for PCHOcM at (■) 10^4 Hz,
(▲) 10^2 Hz, and (●) 1 Hz.
(From ref. [36])

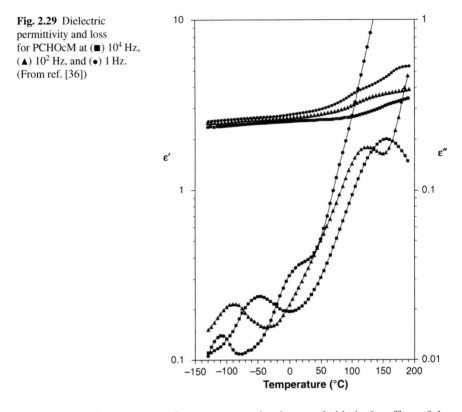

As a general comment on these systems, what is remarkable is the effect of the parity of the number-membered saturated rings. On the other hand, the dielectric as well as mechanical activity is larger in PCOcM in comparison with PCHpM and PCHpMM. Another structural aspect to be taken into account in the viscoelastic behaviour of this family of poly(methacrylate)s is that the insertion of a spacer methylene group enhance the mobility of the side chain giving rise to a new relaxation process and/or a shifting of a pre-existing absorption outside of the experimental range of measurements. Finally a significant differences between the conductive parameters is a characteristic of these systems [36].

Poly(cyclobuty methacrylate)s

Dielectric relaxational behavior

As it was mentioned above, the response of polymers to perturbative dynamic force field involve local motions at short times. These motions are reflected in secondary relaxations commonly known as β, γ, δ etc. Figure 2.30 shows the Arrhenius plot of dieletric α, β and δ relaxations for PCBuMM. At longer times, segmental motions give rise to α relaxation corresponding to the glass-rubber transition and appear at higher temperatures. At very low frequencies, the response to electric force-fields displays the normal mode of relaxation which is strongly dependent on molecular

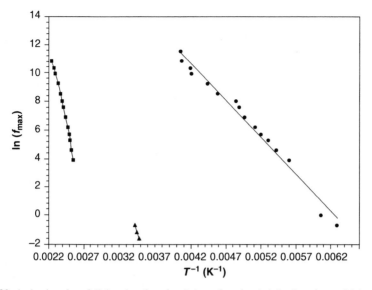

Fig. 2.30 Arrhenius plot of dielectric relaxation (■) α relaxation (▲) β relaxation and (●) γ relaxation. (From ref. [36])

weight. This phenomenon is only present in polymers with significant component of their dipole moment along the local chain axis. Both glass-rubber and secondary relaxations are independent of molecular weight but are also conditioned by the fine structure of polymers. The rigidity of the polymer chain shifts the α relaxation to lower frequencies at a fixed temperature [37]. The interest of study the dielectric behavior of polymers having long side chains with terminal saturated side chains is to the fact that these chains can have several conformational states which can be affected by electric force-field, and therefore to obtain important information about the molecular origin of the molecular motions responsible of the fast relaxation processes. According to Heijboer [28] and Díaz Calleja et al. [36, 94] the number of carbon atoms in the saturated cyclic side chain play an important role on the relaxational behavior. Therefore, beside poly(methacrylate)s containing cyclopentyl, cyclohexyl, cycloheptyl and cyclooctyl rings, polymer with lower number of carbon atoms like cyclobuty methacrylate (CBuM) (See Scheme 2.5) an derivatives are a good contribution to the understanding of the viscoelastic behavior of poly(methacrylate)s.

Figures 2.31 and 2.32 show the dielectric permittivity and loss for poly(cyclobutyl methacrylate) (PCBuM) and poly(cyclobutylmethyl methacrylate) (see Scheme 2.5). In these figures the α relaxation is associated to the glass transition temperature and the β relaxation appear as a shoulder of the α relaxation.

As in the case of other polymers containing saturated cyclic side chains the α relaxation is obscured by low-frequency conductive effects. Moreover in the case of PCBuMM, dielectric activity is also observed. Therefore both polymers show important conductivity contributions at high temperatures and low frequency. The conductivity analysis in this case is also performed using the hopping model

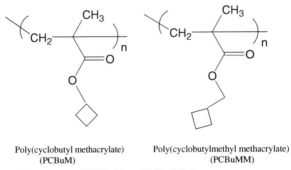

Poly(cyclobutyl methacrylate)
(PCBuM)

Poly(cyclobutylmethyl methacrylate)
(PCBuMM)

Scheme 2.5 Chemical structure of PCBuM and PCBuMM

following the procedure described above for (PtBCHM)s. By this way it is possible to split the conductivity contribution from the dipolar ones. Figure 2.33 represent the plot of $\log \varepsilon''$ versus $\log \omega$ at temperatures above the glass transition. The activation energy for the conductivity, calculated from an Arrhenius plot of $Ln\sigma$ against T^{-1},

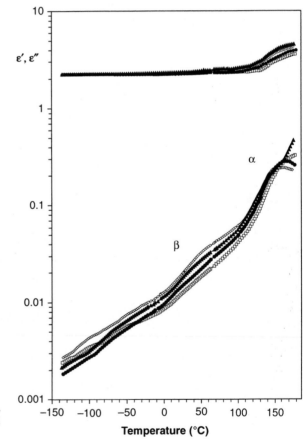

Fig. 2.31 Relative dielectric permittivity and loss for PCBuM (□) 2×10^4 Hz, (●), 2×10^3 Hz, ◊2×10^2 Hz and (▲) 2×10^1. (From ref. [37])

Fig. 2.32 Relative dielectric
permittivity and loss for
PCBuMM (●), 2×10^3 Hz,
◊ 2×10^2 Hz, (▲) 2×10^1 and
(■) 2×10^{-1}. (From ref. [37])

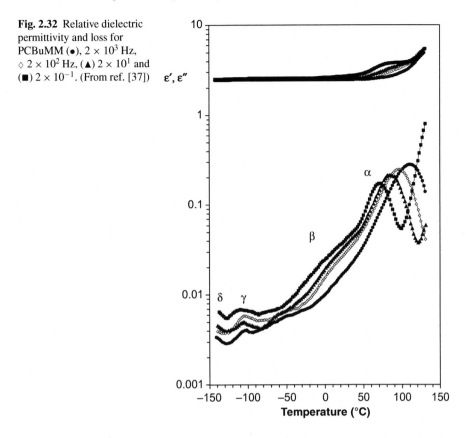

Fig. 2.32 Relative dielectric permittivity and loss for PCBuMM (●), 2×10^3 Hz, ◊ 2×10^2 Hz, (▲) 2×10^1 and (■) 2×10^{-1}. (From ref. [37])

is 32.5 kcalmol^{-1} for PCBuM. An abnormal change in the value of the s parameter, and the conductivity with the temperature has been reported in PCBuMM [37] as can be observed in Fig. 2.34. The parameter s in this figure varies strongly from 0.43 at 110°C to 0.96 at about 170°C. This change in the conductivity is attributed to a different conductive effects depending on the temperature range of measurements what is considered as characteristic of this kind of polymers. For this reason the high-temperature data must be analyzed separatedly. The activation energy for the two conductive phenomena are 24.0 and 63.3 kcal mol^{-1} for low and high temperature zones respectively.

The α relaxation analysis can be performed using the Havriliak-Negami model [70]. The parameters corresponding to this analysis are summarized in Tables 2.9 and 2.10.

On the other hand, as in the analysis of the previous systems the temperature dependence of the α relaxation follow the Vogel-Fulcher-Tamman-Hesse (VFTH) [88–90]. The T_∞ values obtained are 337±5 and 274±5 for PBCHM and PBCHMM respectively. By this way and using equation (2.31) the m' parameter obtained are 1579 and 1804 and the relative free volumes at Tg are 3.2 and 2.6% for PCBuM and PCHBMM, respectively, in good agreement with the free volume theory.

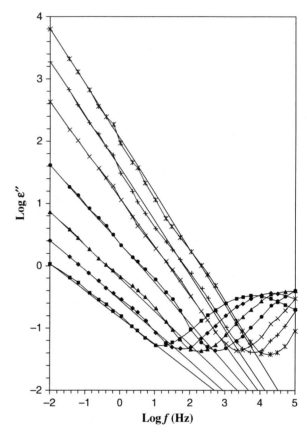

Fig. 2.33 Relative dielectric spectrum(loss permittivity) for PCBuMM. (■) 110°C, (♦) 120°C, (▲) 130°C (●) 140°C, (X) 150°C, (+) 160°C and (*) 170°C. (From ref. [37])

PCBuMM show dielectric activity at very low temperature, as seen in Fig. 2.35.

Two relaxation processes, called δ and γ can be observed, the last as a shoulder of the low temperature side of the β relaxation [37]. Assuming symmetry for these two relaxations the analysis according to Fuoss-Kirkwood empirical equation (2.35) can be performed as in previous systems [36, 37, 69]. Due to the presence of the β relaxation an exhaustive analysis of this systems allow to obtain important information about these relaxation processes.

For this reason in this system he deconvolution of the three relaxations is performed using the addition of two Fuoss-Kirkwood [69] equations for δ and γ relaxations and a power law for the low temperature side of the β relaxation as follows

$$\varepsilon'' = \sum_{i=\gamma,\delta} \varepsilon''^i_{max} \sec h \left[m^i \frac{E_a}{R} \left(\frac{1}{T} - \frac{1}{T^i_{max}} \right) \right] + B \times 10^{\frac{A}{T}} \qquad (2.35)$$

where E_a is the activation energy, R is the gas constant, T_{max} is the temperature where ε'' has a maximum value $\left(\varepsilon''_{max} \right)$, m is a parameter ($0 < m \le 1$) that is

Fig. 2.34 Arrhenius plot for the conductivity (■) and s parameter (▲) of PCBuMM. (From ref. [37])

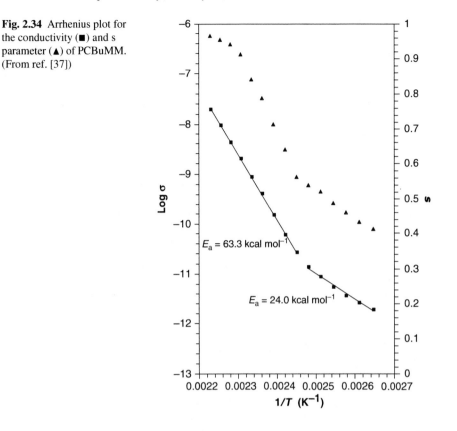

related to the broadness of the peak and A and B are two β relaxation parameters. Figure 2.36 show the deconvolution for δ, λ and β relaxations at 1000 Hz [37].

In conclusion in these two polymeric systems, i.e. PCBuM and PCBuMM, a prominent α relaxation is observed but overlapped by conductivity contributions. The effect of a spacer group as in other poly(methacrylate)s contribute to

Table 2.9 Parameters of the Havriliak-Negami equation for PCBuM. (From ref. [37])

T (°C)	ε_{U_α}	$\Delta\varepsilon_\alpha$	τ_α (S)	$\bar{\alpha}$	$\bar{\beta}$
152	2.58	2.30	4.34×10^{-4}	0.29	0.81
156	2.67	2.21	1.46×10^{-4}	0.32	0.92

Table 2.10 Parameters of Havriliak-Negami equation for PCBuMM. (From ref. [37])

T (°C)	ε_{U_α}	$\Delta\varepsilon_\alpha$	τ_α (S)	$\bar{\alpha}$	$\bar{\beta}$
75	2.32	1.77	3.06×10^{-2}	0.36	0.72
80	2.35	1.67	0.58×10^{-2}	0.39	0.80
85	2.36	1.58	0.19×10^{-2}	0.43	0.82

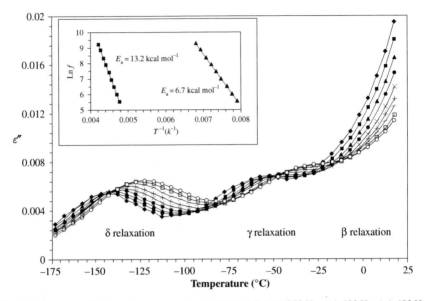

Fig. 2.35 Loss permittivity at low temperature for PCBuMM (♦) 253 Hz, (■) 403 Hz, (▲) 683 Hz, (●) 1020 Hz, (X) 1620 Hz, (+) 2610 Hz, (−) 4000 Hz, (□) 6670 Hz, (○) 10000 Hz. Insert: Arrhenius plot for the γ relaxation (■) and δ relaxation (▲) for PCBuMM. (From ref. [37])

a decreasing of Tg from PCBuM to PCBuMM. The activation energies of the γ relaxation are comparable to the flipping motions of cyclohexyl and cyclooctyl polymers [32, 33, 36] and with other polymers containing cyclobutyl groups in the side chain. According to the analysis of these two polymers it can be observed

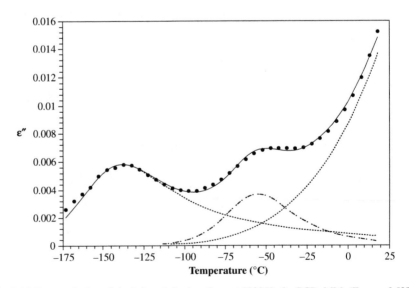

Fig. 2.36 Deconvolution of the δ, λ and β relaxations at 1020 Hz for PCBuMM. (From ref. [37])

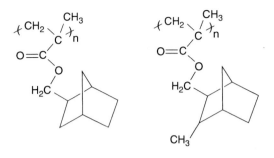

P2NBM   P3M2NBM

Scheme 2.6 Poly(2-norbornyl methacrylate) (P2NBM), and poly(3-methyl-2-norbornyl methacrylate) (P3M2NBM). (From ref. [35])

that the higher flexibility of methylcyclohexyl groups enhances dielectric activity in comparison with polymers containing cyclobutyl groups. Polymers having even saturated rings in the side-chain have more sub-glass dielectric activity than the odd membered. However, the activity of cyclobutyl group is less than that in the case of cyclohexyl [28] and cyclooctyl [36] substituent, and the relaxations are observed at lower temperatures [37].

2.4.2.4 Poly(methacrylate)s Containing Norbornyl Groups

Dielectric relaxational behavior
Another interesting family of saturated cyclic poly(methacrylate)s are those containing norbornyl groups which show a very interesting behavior from viscoelastic point of view. Scheme 2.6 show these structures in which as the previous cases an spacer group have been inserted in order to get confidence about the effect of small structural changes on the viscoelastic responses.

Figures 2.37 and 2.38, show the isochronal curves of the permittivity and loss factor for P2NBM and P3M2NBM as a function of temperature at fixed frequencies. A prominent relaxation associated with the dynamic glass transition is observed in both polymers. Clearly the effect of the methyl substitution in position 3 of the norbornyl group is to decrease the temperature of this relaxational process.

At low frequencies, conductive phenomena overlap the loss factor. Owing to the fact that these conductive effects also affects the real part of the complex permittivity, one can conclude that these effects are not only due to free charges conduction but also a blocking electrode process is present. For instance, Fig. 2.39 shows the loss permittivity of P3M2NBM in the frequency domain at 105°C. At this temperature, secondary processes are absent in the range of frequencies shown in this figure. In this case in order to remove the conductive effects be a straight line with slope -1 in a double logarithmic plot of log ε'' vs log f, is used. As in the previous cases the α relaxation can be analyzed using the Havriliak-Negami procedure [35, 70].

Fig. 2.37 Main Figure: dielectric permittivity and loss for P2NBM (■) 10^4 Hz, (▲) 10^2 Hz, (●) 1 Hz. Insert, showing the low temperature process: (■) 5×10^4 Hz, (▲) 3×10^4 Hz, (x) 2×10^4 Hz, (*) 10^4 Hz, (●), 5×10^3 Hz and (−) 2×10^3 Hz. (From ref. [35])

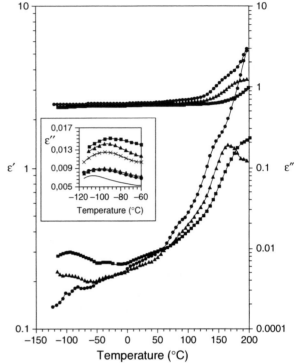

The results reported for these systems, indicate that the subglass relaxation in P3M2NBM is indeed a broad and weak relaxational process [95]. The low strength of the observed secondary processes in these systems, has been attributed to the bulkiness of the side chain. In one case, the low strength of the secondary relaxation prevents the analysis of this peak in terms of an empirical relaxation. The effect of the methyl substituent on the norbornyl ring is to lower the position of the value of the peak of the α-relaxation about 50 K. Another effect that can be observed in the low frequency side of the spectra in both polymers is the conductive one.

2.4.3 Poly(methacrylate)s Containing Heterocyclic Side Groups

2.4.3.1 Poly(tetrahydropyranyl methacrylate)

(a) Dielectric relaxational behavior

Although molecular mobility is severely restricted below the glass transition temperature, the dynamic glass transition temperature (main transition or, conventionally -relaxation) in polymers as it have been described above, is usually accompanied by subglass secondary relaxations labeled as β, γ, δ, relaxations. The glass transition at low temperatures is assumed to be caused by the cooperative motion of many particles, while the secondary relaxations have a more localized molecular

Fig. 2.38 Main Figure: dielectric permittivity and loss for P3M2NBM (■) 10^4 Hz, (▲) 10^2 Hz, (●) 1 Hz. Insert, showing the low temperature process: (■) 5×10^4 Hz, (▲) 3×10^4 Hz, (x) 2×10^4 Hz, (*) 10^4 Hz, (●), 5×10^3 Hz, (+) 3×10^3 Hz, (−) 2×10^3 Hz and (−) 10^3 Hz. (From ref. [35])

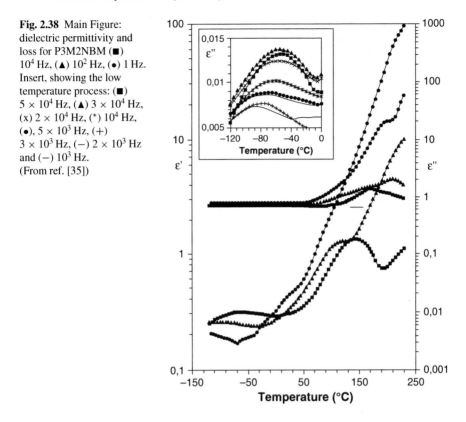

origin. The number of these secondary relaxations usually depends on the nature of the side chain. Thus, polymers with side chains with large number of degrees of freedom are susceptible to undergo several molecular motions, which are visualized as a large number of secondary relaxation processes. Particularly, polymers containing cyclohexyl side groups, as it was described above (see Scheme 2.7), are characterized for displaying considerable dielectric and mechanical activity in the glassy state [27, 28, 32, 45, 48, 51, 82, 83, 96–103]. Therefore, this kind of polymers is specially suitable to study the lateral chain dynamics above and below the glass transition temperature.

On the other hand the effect of small modifications of the side chain affect the dielectric response what have been reported in other polymer systems containing cyclohexyl groups [27]. In this case the substitution of carbon atoms by oxygen in the cycloheyl ring affect the relaxational behavior of the resulting polymer.

(b) Molecular dynamic simulation

The dielectric and mechanical relaxations on poly(1,3-dioxan-5yl-methacrylate) (PDMA) [104], show that this polymer present a variety of absorptions due to the versatility of its structural moiety [105]. Recently this behavior have been studied by molecular dynamic simulation using different methods and force fields [106–109]. These polymers are analyzed from molecular simulation using different ways but

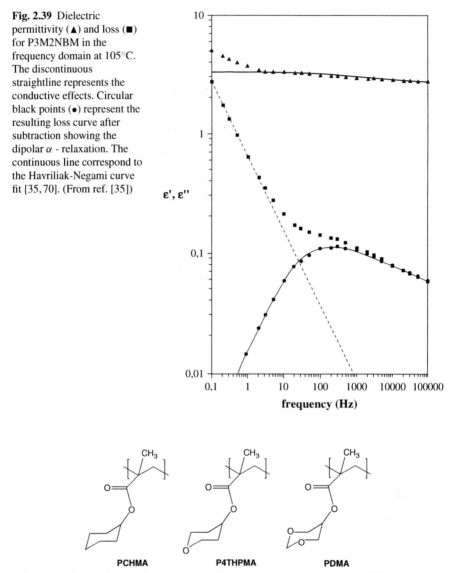

Fig. 2.39 Dielectric permittivity (▲) and loss (■) for P3M2NBM in the frequency domain at 105°C. The discontinuous straightline represents the conductive effects. Circular black points (●) represent the resulting loss curve after subtraction showing the dipolar α - relaxation. The continuous line correspond to the Havriliak-Negami curve fit [35, 70]. (From ref. [35])

Scheme 2.7 Schematic chemical structures of Poly(cylohexyl methacrylate) (PCHMA), Poly(4-tetrahydropyranyl methacrylate) (P4THPMA) and poly(1,3-dioxan-5-yl-methacrylate) (PDMA). (From ref. [38])

other procedures as ab initio molecular orbital calculations should also be useful for the same purpose. It is interesting to analyze the conformational behavior of these polymers from molecular simulation point of view. These systems were studied using a PC-MODEL, software [110], which is based on force field, called MMX that is derived from MM2(P) [111].

The use of this model compounds is advisable to simplify the calculations and the corresponding interpretations and to get confidence about the relaxational behavior of these systems which can be obtained in a similar way to those of other authors [27, 99–102, 112–119]. This empirical field force includes intermolecular as well as intramolecular contributions. The nonbonded energy function expresses interactions between atoms that are not bonded to each other. It is splitted into a van der Waals steric component and an electrostatic component, dealing with interactions between charges and dipoles. In this analysis, the intramolecular energy function is splitted into a connectivity term, the bond stretching function and flexibility terms, the angle bending and the torsional functions, as well as cross-terms describing the coupling of stretch–bend, bend-bend, and torsional–stretch interactions that are taken into account. The Wilson (E_{OOP}, umbrella out-of plane) term has also been included, but their contribution to the global energy was small in all cases.

Figure 2.40 show the loss permittivity of P4THPMA where three subglass absorptions, labeled as δ, γ and β relaxations, centered at $\sim$140, $\sim$190, and $\sim$260 K at 1 Hz respectively, followed in the increasing order of temperatures by glass-rubber process or α relaxation located at $\sim$420 K at the same frequency.

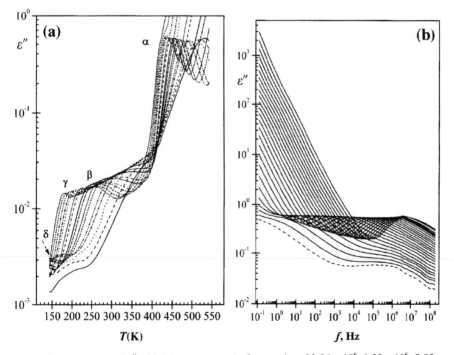

Fig. 2.40 Dependence of ε'' with (**a**) temperature (at frequencies of 2.96×10^6, 1.32×10^6, 5.85×10^5, 1.16×10^5, 5.14×10^4, 1.02×10^4, 4.51×10^3, 1.34×10^3, 5.94×10^2, 1.17×10^2, 5.52×10^1, 1.03×10^1, 4.58×10^0, 1.36×10^0, 4.02×10^{-1} Hz) and (**b**) frequency at temperatures from 418 to 543, step 5K) for P4THPMA. (From ref. [38])

Fig. 2.41 Dependence
of log f_{max} with the inverse
of temperature in the range of
α ($\triangle$ P4THPMA), β ($\square$
P4THPMA), γ (P4THPMA,
$\bigcirc$PCHMA, PDMA) and δ
($\lozenge$PTHPMA) relaxations.
(From ref. [38])

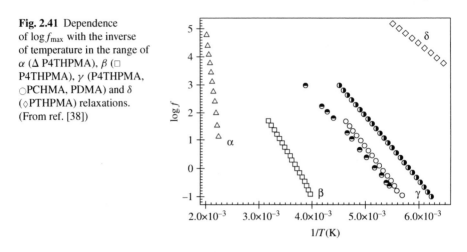

At high temperatures and low frequencies conductivity contribution are important since the loss permittivity tends to increase continuously. Figure 2.41 show the Arrhenius plot for the determination of the activation energy for the δ relaxation which is about $\sim$28 kJ mol^{-1}. This is a value very close to those reported for similar structurally poly(methacrylate)s [28, 29]. Increasing the temperature, a γ relaxation is observed.

This process is partially overlapped with the next process, the β relaxation. To analyze the loss permittivity in the subglass zone in a more detailed way, the fitting of the loss factor permittivity by means of usual equations is a good way to get confidence about this process [69]. Following procedures described above Fig. 2.42 represent the lost factor data and deconvolution in two Fuoss Kirwood [69] as function of temperature at 10.3 Hz for P4THPMA. In Fig. 2.43 show the γ and β relaxations that result from the application of the multiple nonlinear regression analysis to the loss factor against temperature. The sum of the two calculated relaxations is very close to that in the experimental curve.

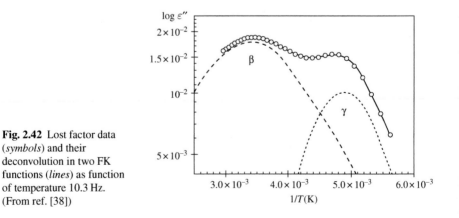

Fig. 2.42 Lost factor data
(*symbols*) and their
deconvolution in two FK
functions (*lines*) as function
of temperature 10.3 Hz.
(From ref. [38])

The evolution of the frequency of the maxima show a linear dependency with the inverse temperature according to the expectations as shown in Fig. 2.41. The activation energies calculated from the Arrhenius plot for each process are 68.2 ± 0.6 kJ mol^{-1} for β relaxation and 48.0 ± 0.3 kJ mol^{-1} for γ relaxation. These values are very close to those found for structurally related to P4THPMA such a PCHM [29, 30], PMCHMA [59], poly(2-chlorocyclohexyl methacrylate) [33] and PDMA [104] (70–82 and 41–50 kJ mol^{-1} [38]).

It is convenient to remember that the γ-relaxation in polymers containing cyclohexyl groups was assigned to the change in the conformational chair-to inverse chair in the cyclohexyl group [28, 30, 51, 97–99, 104, 114, 120]. On the other hand, the β relaxation in PMMA is attributed by several authors [114–120] to the partial rotation of the lateral ester group as a whole coupled with some type of motions of the main chain. By using 2D-NMR measurements, Spiess and coworkers [121] have concluded that for PMMA and PEMA, the unusual main chain mobility below Tg is coupled to the β relaxation process, which involves $180°$ flips of the carbonyl side groups.

As usual, the relaxation strength of the subglass processes increases as the frequency decreases. The relaxation strength of the γ relaxation in P4THPMA is close to the corresponding to PCHMA [30] and significantly lower than that of the corresponding to PDMA [38, 104].

As can be seen in this system at high temperatures and low frequencies, the α relaxation is contaminated by conductivity contributions (combination of ionic conduction and interfacial polarization of ionic conduction and interfacial polarization of electrodes), which are observed through the continuous increase of the loss permittivity. This phenomenon suggests that conductivity contributions are dominant over dipolar processes in this zone [38]. Using Havriliak-Negami procedure [70] is possible to analyze the α relaxation. As a general result in these systems it is found that the relaxation strength of the α process slightly decreases when the temperature increases. The values obtained for this parameter are higher than that corresponding to PCHMA [30] and poly(2-, 3-, and 4-methylcyclohexyl methacrylate)s [51], suggesting that the mean square dipole moment of the polymer under study is higher than that of PCHMA and PMCHMA. The substitution of one carbon for one more polar group such as oxygen could be responsible of such an effect [88–90].

The temperature dependence of the α relaxation may be analyzed in the context of the free volume theory by means of the Vogel-Fulcher Tamman-Hesse (VFTH) [88–90] procedure. The temperature dependence of the α relaxation with the frequency is shown in Fig. 2.43. The obtained value for the free volume is $\sim 3.1\%$: this is a lower value than those reported for other related polymers. i.e. P2MCHMA ($\sim 3.6\%$, P3MCHMA ($\sim 4\%$), P4MCHMA ($\sim 5\%$) and PDMA ($\sim 3\%$), but somewhat higher than that predicted by the free volume theory (2.5%) [122]. These facts are probably due to the bulkiness of the side chain groups, which is higher for P2MCHMA, P3MCHMA, P4MCHMA and PDMA in comparison with that of PCHMA and P4THPMA.

A powerfull tool to provide an explanation of the molecular origin of the observed secondary relaxation is molecular simulation technique, to elucidate the

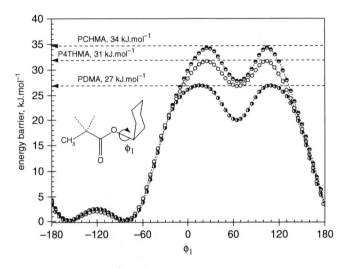

Fig. 2.43 Potential energy (kJ mol^{-1}) profile for rotation of O-C bond for (ϕ_1: $-180°$ to $+180°$) for one-unit model compounds of (◐) PCHMA, (○) P4THPMA and (◑) PDMA. (From ref. [38])

specific motions responsible of the observed secondary relaxation, the calculation and comparison of the conformational barriers with the activation energies obtained from dielectric measurements.

Scheme 2.8 shows the optimization geometry of the one-unit model compound of P4THPMA.

The activation energy obtained by dielectric measurements for δ –process is ~28 kJ mol^{-1}. This value is close to that obtained for PCHMA [30, 38]. Therefore the molecular origin of both relaxation processes should be related. This relaxation is observed in cyclohexyl compounds at temperatures below the γ relaxation and, several authors [125] have related this relaxation to the motions of cyclohexyl group as a whole. In order to interpret the molecular origin of the δ -transition, calculations

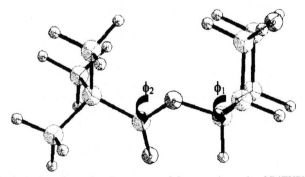

Scheme 2.8 Optimized conformational geometry of the repeating unit of P4THPMA (From ref. [38])

confined to the rotation of the cyclohexyl ring around the O-C bond allow to get confidence about this transition.

Figure 2.43, taken from ref. [38], show the potential energy profiles obtained for the rotation of the O-C bond.

In this calculation the specified angle (ϕ_1) is constrained to certain desired values (in the range $-180°$ and $180°$, step $5°$) and the remaining parts of the molecule are allowed to move freely to minimize the energy. The energy barrier is of $\sim 31 \text{ kJ mol}^{-1}$. This value is close to that evaluated from dielectric measurements ($\sim 28 \text{ kJ mol}^{-1}$) [38].

As temperature increases another relaxation processes labeled as γ-relaxation is observed and is associated with the barrier hindering the chair-to-chair interconversion of the cyclohexyl group. It is known that the chair-to-chair interconversion is a complex process produced by successive conformational changes within the molecule [38]. When an isolated cyclohexyl ring undergoes internal rotations, the chair-to-chair transition in the cyclohexyl ring follows the sequence given by chair→half-chair→boat-half-chair→ chair. The half-chair is not stable because the hydrogen atoms at the foot are eclipsed with those on the adjacent carbons, and because of this fact, there is some angular strain. This angular strain occur because the molecule is nearly planar and are destabilized by torsional strain. When the cyclohexyl ring is bounded to the polymer chain, the motion is something different to that before mentioned in this case, the motions that produces the transition from one chair form to another fixes the oxycarbonyl group to the main chain and therefore this group remains nearly in the same position above and after the chair-to-chair transition. The energy barrier in this case is calculated as the difference between the energy of the less favorable conformation (the half-chair) and the more stable conformation (chair). The energy barrier obtained with a simple model compound of one unit, 46.1 kJ mol^{-1}, is close to that evaluated by dielectric measurements, 48 kJ mol^{-1}. Therefore the molecular mechanic calculations strongly support the idea of intramolecular barrier for the γ relaxation associated with the chair-to-chair interconversion of the cyclohexyl ring at the end of the side chain.

PTHPMA is a good example of the analysis of relaxational processes in polymers containing saturated and substituted side rings, and, allow to an understanding the origin of the molecular motions responsible of the fast relaxational complex processes in these systems.

2.4.3.2 Poly(tetrahydrofurfuryl methacrylate)

Polymers containing heterocyclic rings in the side chain have good biocompatibility, and they have found numerous medical applications. They are potential candidates [64, 123–128], as well as in medical implants, as materials for cartilage repairs [64, 129–133] and in other medical products [64]. In all these applications the diffusion of water in the polymer matrix is of fundamental importance: it presence controls the properties of the swollen polymers [36, 123, 133–136]. The glass transition temperature is lowered by the presence of adsorbed water in the rubbery regions, resulting in a significant increase in the segmental mobility of the polymer chains. So, the diffusion of body fluids (water) allows the drug molecules to diffuse out of the polymer. The water diffusion as well as the effect of the addition of an antibiotic has been studied by Downes and coworkers [126, 127]. Braden and coworkers [123–126, 131, 134, 136–138], have shown that homo and copolymers containing tetrahydrofurfuryl methacrylate (PTHFM) are very promising biomaterials for use in the field of medicine and dentistry [64]. For dental applications this polymer is used as comonomer with ethyl methacrylate, hydroxyethyl methacrylate and bisphenolA-glycidyl methacrylate with up to 30% of THFM by weight. These authors also shows [137] that PTHFM posses some unique characteristics with respect to its biocompatibility and behavior in water. The water uptake is high (>70%) and very slow (over three years) but the material remain rigid throughout the process; that is, does not form a hydrogel [64, 123, 137]. It has a biological tolerance by dental pulp [139] and is superior to other glassy poly(methacrylate)s in terms of drug delivery [64]. Much effort has been devoted to studying the organization and properties of water in this type of material, and its effect on their properties. A variety of experimental techniques have been employed in order to clarify that behavior [64].

Analogous poly(itaconate)s polymers like poly(ditetrahydrofurfuryl methacrylate) have been also studied because there are at least two advantages in using itaconate acid based polymers over methacrylate acid derivatives: itaconic acid can be obtained through fermentation from renewable, non petrochemical sources and the toxicity of its derivatives is lower than for methacrylate derivatives [64, 140].

Because of the potential interest of these materials, Sanchis and coworkers [64] shows that it is interesting to know the dielectric behavior of two heterocyclic poly(methacrylate)s: PTHFM and poly(3-methyl tetrahydrofurfuryl methacrylate) P3MTHFM, (see Scheme 2.9)and the comparison of the relaxation properties of these two polymers.

As it have been demonstrated above, dielectric spectroscopy yields a wealth of information on the different molecular motions and relaxations processes, which are

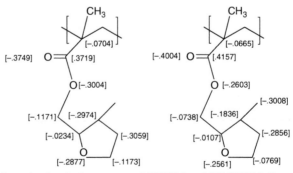

Scheme 2.9 Schematic chemical structures of PTHFM and P3MTHFM. Between squares the dipoles on the one-unit model on the two polymers evaluated using PM3 (converge limit 0.01) for energy minimization and Molecular Mechanic (MM +) force field for molecular dynamic at 300 K. (From ref. [64])

useful in order to understand the behavior of these material in biomedical applications [64].

(a) Dielectric relaxational behavior

Figure 2.44a, b show the dielectric loops at several frequencies in the full range of temperatures. These figures show together with the α relaxation located at $\sim$340–350 K at 10^0 Hz, conductive effects and two prominent subglass absorptions, labeled as δ and γ -relaxations centered at $\sim$120 K and $\sim$160 K at 10^0 Hz respectively. A shoulder reminiscent of the β relaxation can be observed close to the room temperature. In this case as in other previously shown, it is evident that the α relaxation cannot be easily observed due to the continuous increase of the loss permittivity. This phenomenon, that is the continuous increase of the loss permittivity with increasing temperatures, suggests that conductivity contributions are dominant in this region over dipolar processes [64].

Figure 2.44c, d shows the dielectric relaxation spectra of PTHFM and P3MTHFM taken from ref. [64]. In this case, it is possible to observe that loss permittivity increases linearly with decreasing frequency, at low frequencies. At high temperatures, and for both polymers a linear behavior with slope close to minus one is observed when data are represented in a log-log plot. Moreover a change in the slope of the linear behavior is also observed in the case of PTHFM, at $\sim$$10^0$ Hz, and 423 K. This anomalous dielectric behavior is also known as low frequency dispersion (LFD). Its occurs primarily in dielectric materials with large densities of low mobility charge carriers [64, 141]. This process is distinguished from dc-conductivity and from electrode polarization effects, which is due to the interaction of sample with the electrode interface, by the fact that in the frequency range where the relaxation take place, the permittivity tend to be parallel to the imaginary permittivity.

According to these results, the conductivity and other possible contributions to the loss permittivity are more pronounced for PTHFM than for P3THFM. This phenomenon is associated to the residual water in the samples that cannot be removed after drying in vacuo at room temperature. This is an important trouble in the manipulation of polymeric materials, because it is very difficult to eliminate water and in the case of hydrophilic polymers the trouble is worst. These kind of materials are hygroscopic and the dryness condition under the experiments are performed must be strictly careful.

There are several works dealing with the behavior of polymers with similar structures in the sense that these materials are able to form clusters by reorganization of dipolar groups. To form clusters some degree of freedom and chain mobility is necessary. According to Riggs [136], there are at least to ways in which clustering can occur that distinguish, PTHFM from the other poly(methacrylate)s: ring opening of the tetrahydrofurfuryl ring or clustering of the actual tetrahydrofurfuryl ring itself [64]. Therefore when PTHFM absorbs water, it form clusters probably associated to reorganization of tetrahydrofurfuryl rings. Although the exact mechanism for cluster formation is still unclear, reorientation of the rings could be involved in the initial stage of cluster formation. Scheme 2.9 shows that the rings has a strong molecular dipole associated with it. The dipole of the carbonyl bond is the largest

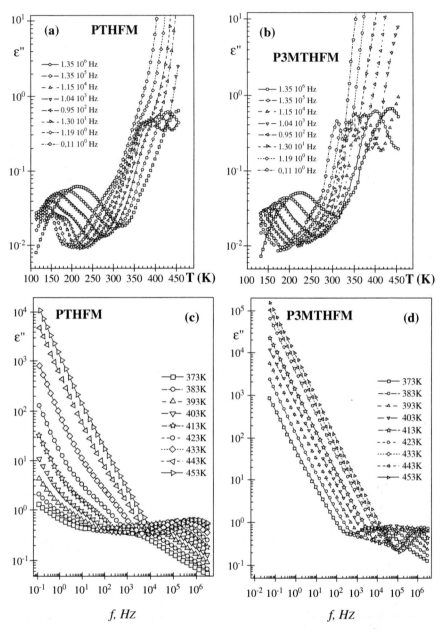

Fig. 2.44 Dependence of ε'' with temperature and frequency for PTHFM (**a,c**) and P3THFM (**b,c**). (From ref. [64])

one but it is in close proximity to the polymer backbone, so it will interfere with water absorption. The oxygen of the tetrahydrofurfuryl ring is more readily available, and enjoys a less conformationally restricted environment because of its ability to rotate about the CH_2 unit between the carbonyl group and the tetrahydrofurfuryl group [64]. On the other hand in the case of P3THFM in which a methyl group is inserted in position 3, could hinder the ring reorientation and should prevent cluster formation. Scheme 2.9 Show the calculated dipoles of the polymers which demonstrate that the dipole moment of the substituted polymer is higher in comparison with the unsubstituted one [64].

The higher complexity of the dielectric spectra of PTHFM and P3MTHFM, could be responsible for the presence of two mechanism of water sorption: (a) the sorption of water by the polymer matrix and (b) clustering with tetrahydrofurfuryl group. A cluster is defines as a group of hydrogen-bonded water molecules giving rise to a network through which the proton transport take place [64]. Water cluster systems have also been observed in many partially hydrated biological and pharmaceutical systems [142–144]. On the other hand, Dissado and Hill [64, 145], described the mechanism of this type of anomalous dielectric response, associated to the presence of water, in terms of charge transport (proton hopping) both within and between clusters os associated charge carriers, the percolation cluster model.

Sanchis and coworker [64] in order to insight something about this fact and in order to get confidence about this phenomenon, have used the electric modulus formalism [146], ($M^* = 1/\varepsilon^*$) to represent the experimental data. The advantages of this kind of representation are evident due to the better resolution observed for dipolar and conductive processes. The imaginary part of M^* as a function of frequency at 423 K, for both polymers, is shown in Fig. 2.45. The curve corresponding to PTHFM shows a complex behavior at low frequencies, which presumably is the result of the superposition of the two conductive processes.

In addition, the half-widths of the loss modulus curve as a function of the frequency are higher than that corresponding to the P3MTHFM. For the last polymer,

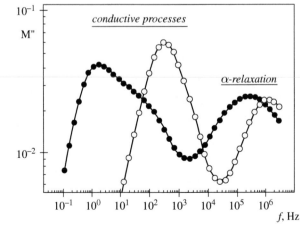

Fig. 2.45 Frequency dependence of M" for PTHFM (*full symbol*) and P3MTHFM (*open symbol*) at 423 K. (From ref. [64])

the half-width of the loss modulus is near 1.144, as corresponding to a process governed by a single Debye peaks. Owing to the fact that pure conductive phenomena, are usually well described by a single Debye peak, the higher values found for the half-width of the process observed in PTHFM, suggest that more than one conductive process are present.

In this case in order to characterize the dipolar and conductive contribution at high temperatures, Sanchis and coworkers [64] have used the following equation:

$$\varepsilon''(\omega) = \varepsilon''_{cond}(\omega) + \varepsilon''_{dipolar}(\omega) \tag{2.36}$$

Where $\varepsilon''_{cond}(\omega)$ and $\varepsilon''_{dipolar}(\omega)$, represent the conductive contribution to the loss permittivity, $\varepsilon''_{cond}(\omega)$, is given by [146–149]:

$$\varepsilon''_{cond}(\omega) = \left(\frac{\sigma}{\varepsilon_{vac} \cdot \omega}\right)^s \tag{2.37}$$

where σ is the conductivity, s $\leq$ 1, and ε_{vac} is the vacuum permittivity (= 8.854 pFm^{-1}).

As in the systems described above the α relaxation can be modeled by using the classical Havriliak Negami [70, 87] procedure.

The s parameter following this procedure is found to be between $\sim$0.91 and 0.95 and the conductivity increases $\times 10^{-8}$ S cm^{-1}. The activation energy, for this conductive process, obtained from the Arrhenius plot was equal to 100.5 kJ mol^{-1} (1.04 eV). As usual, the dielectric strength of the α -relaxation $\Delta\varepsilon = \varepsilon_{0\alpha} - \phi_{\infty\alpha}$, decreases when the temperature increases. The shape parameter for both parameters are nearly temperature independent.

To give account for the second conductive process it is necessary to modify the model, and therefore the conductive contribution to the loss permittivity in 2.47 can be characterized by:

$$\varepsilon''_{cond}(\omega) = \left(\frac{\alpha}{\varepsilon_{vac} \cdot \omega}\right)^s + r^{-\frac{\nu}{2}} \cdot (\varepsilon_0 - \varepsilon_\infty) \cdot \sin \nu\theta \tag{2.38}$$

Following this procedure it is possible to observe that the s parameter takes values close to the unity and σ increases with increasing temperature from 3.62×10^{-12} to 6.11×10^{-11} S cm^{-1}. The second conductive process is symmetric because the ν parameter is as unit at all temperatures. This type of symmetric process have been observed in different materials where the water is present [143, 150]. The activation energy obtained from Arrhenius plots for these conductive processes are 118 kJ mol^{-1} (1.23 eV) and 61.5 kJ mol^{-1} (0.64 eV). The activation energies related with the symmetric conductive process are similar to those observed in biological materials where the presence of the water clusters had been demonstrated [64, 151]. The magnitude of this activation energy be explained by the way in which the water interacts with the amorphous matrix.

Figure 2.46 is an example of the frequency dependence of ε'' for PTHFM at 393 K. The sum of the three calculate processes, continuous line, is very close to

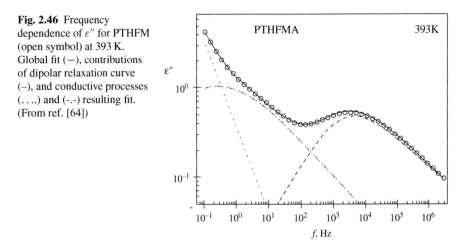

Fig. 2.46 Frequency dependence of ε'' for PTHFM (open symbol) at 393 K. Global fit (−), contributions of dipolar relaxation curve (−), and conductive processes (....) and (-.-) resulting fit. (From ref. [64])

the experimental data, open symbol. In this figure [64] broken lines represent the result of the according to equations (2.36) and (2.38). Despite of the complexity of the experimental spectra, the deconvolution represented in Fig. 2.46 is a convenient approach to interpret the behavior of this polymer.

It is possible to observe that the dielectric strength of PTHFM is slightly lower than P3THFM. This suggests that the mean square dipole moment is lower in PTHFM, than for P3THFM.

The temperature dependence of the α relaxation analyzed using the Vogel-Fulcher Tamman-Hesse (VFTH) equation (2.31) [88–90]. The temperature dependence with the frequency of the α relaxation is shown in Fig. 2.47

Values of the free volume are 3.19% and 3.76% for PTHFM and P3MTHFM respectively. These values are similar to those commonly reported for amorphous polymers [122].

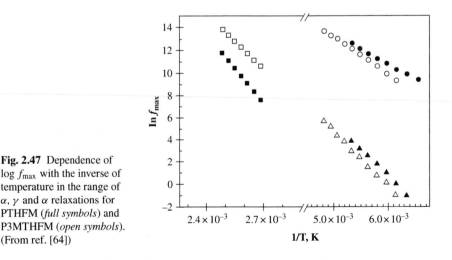

Fig. 2.47 Dependence of log f_{max} with the inverse of temperature in the range of α, γ and α relaxations for PTHFM (*full symbols*) and P3MTHFM (*open symbols*). (From ref. [64])

The analysis in the subglass region allow to fit the loss factor permittivity by empirical equations. As in systems previously analyzed a reliable model to represent secondary relaxation is that of Fuoss-Kirkwood [9]. Assuming that the two overlapped contributions for δ and γ relaxation are additive Sanchis and coworkers [64] have proposed the following equation:

$$\varepsilon'' = \sum_{i=\gamma,\delta} \varepsilon''_{max,i} \cdot sech\,(m_i \cdot x)\,\text{with}\,x = \text{Ln}\frac{f}{f_{max,i}} \qquad (2.39)$$

where m_i is an empirical parameter lying in the interval $0 < m \le 1$, f_{max} is the frequency associated with the maximum of the isotherms. The m parameter defines the broadness of the relaxation in such a way that the higher m is, the narrower the distribution. Particularly for a Debye type relaxation, m = 1. Figure 2.48 taken from ref. [64], show an example of the fitting, where, the frequency dependence of ε'' at 168 K P3MTHFM. The m_γ and m_δ values increase with increasing temperature. The low values obtained for these parameters are indicative of the distributed character of the processes. Another interpretation of these results is associated with the fact that the strength of the subglass relaxation for PTHFM and P3MTHFM show a weak temperature dependence. In fact, slight increase with decreasing temperature is observed in Fig. 2.47. Clearly the inclusion of a methylene group in the ring reduce the dielectric strength of the γ relaxation, and slightly increases the dielectric strength of the δ process. The activation energies obtained from the Arrhenius

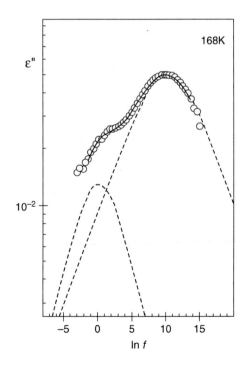

Fig. 2.48 Deconvolution of dielectric loss in the frequency domain for P3MTHFM at 168 K. *Open circles* represent the experimental losss factor, *dashed lines* the deconcolution processes, and continues lines represent the sum of both deconvolutes processes.(From ref. [64])

plots represented in Fig. 2.47, in this case are 21.2 ± 0.2 kJ mol^{-1} for δ relaxation, 39.6 ± 0.9 kJ mol^{-1} for γ relaxation for PTHFM and 27.0 ± 0.4 kJ mol^{-1} for δ relaxation, 44.1 ± 0.8 kJ mol^{-1} for γ relaxation in P3MTHFM. According to these results the activation energy for these relaxations in PTHFM and P3MTHFM is scarcely affected by the substitution of one hydrogen with a CH$_3$- in tetrahydro-furfuryl ring. Nevertheless the, activation energy of the δ -relaxation is appreciably affected by such substitution [64].

(b) Molecular Mechanics simulations

The interpretation of the molecular origin of the observed secondary relaxations as in the previous cases can be simulated by molecular dynamic simulations. This can be performed by comparison of the conformational energies barriers with the activation energies (E$_a$) obtained from dielectric measurements [64]. Scheme 2.10 represents the optimized geometry of a one-unit model (monomer) compound of the two polymers, calculated by using the force field MMX is represented in Scheme 2.10. From DRS experiments, the activation energy values for PTHFM is $\sim$21 kJ mol^{-1}, and $\sim$27 kJ mol^{-1} for P3MTHFM for the δ process. These values are very close to those obtained by MM calculations i.e. 19 and 21 kJ mol^{-1}, when de dihedral angle (ϕ_1) is constrained to certain desired values in the range $-180°$ and $180°$, step $5°$, and the remaining parts of the molecule are allowed to move freely to minimize the energy. In the same way when the dihedral angle (ϕ_1) is constrained, energy barriers of $\sim$36 kJ mol^{-1} for PTHFM and $\sim$41 kJ mol^{-1} for P3MTHFM. As can been reported these values are very close to those obtained DRS experiments. i.e. $\sim$40 and $\sim$41 kJ mol^{-1}. respectively.

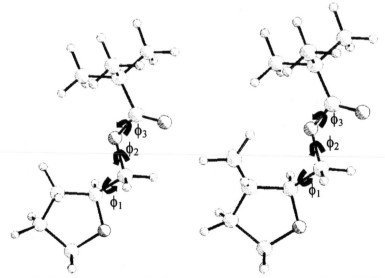

Scheme 2.10 Optimized conformational geometry of the repeating unit of PTHFM and P3MTHFM

According to the results shown for this polymer, it is possible to confirm the hypothesis of Riggs et al. [129] concerning the existence of two mechanisms of water absorption. On the other hand it is interesting to note that the dielectric analysis of these polymers allow to know the importance of water sorption in this kind of polymers what is very important from technological an medical point of view. Therefore dielectric measurements on these kind of polymers result in a powerfull tool to analyze the effect of water absorption on the polymeric matrix and then to applications of these materials.

2.4.4 Poly(methacrylate)s Containing Aromatic Side Chains

2.4.4.1 Poly(dimethylphenyl methacrylate)s

(a) Dielectric relaxational behavior

Whereas it is well recognized that the most prominent relaxation in amorphous polymers, the glass-rubber or α relaxation, is caused by long-range, generalized and cooperative motions of the main chain [152] the exact nature of the secondary relaxation taking place at temperatures below the glass transition temperature, where long-range motions are frozen, is not well understood [153].

As the subglass relaxation or β process is related to reorientation of flexible side groups in polymer chains, it was considered interesting to investigate a group of synthetic poly(dimethyl substituted) polymers, such as poly(dimethylphenyl methacrylate)s (see Scheme 2.11). In this way, it should be possible to obtain a better understanding of the molecular origin of the relaxation process and to estimate the influence of the steric hindrance and dipole-dipole interactions relative to the positions of the two methyl groups which both lie in a single plane 120° or 180° of the aryl ring. To improve the study of these systems, in this case another experimental technique is used by Díaz Calleja et al. [42] i.e. thermally stimulated current (t.s.c.) measurements. Nevertheless conventional alternating current (a.c.) dielectric techniques are also studied because when both techniques are used together, cover a wide frequency range [155]. t.s.c. is a highly sensitive technique

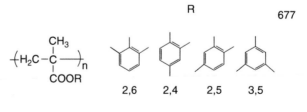

R   677

2,6   2,4   2,5   3,5

PDMP

Scheme 2.11 Chemical structures of poly(dimethylpheyl methacrylate)s.Poly(2,6-dimethylphenyl methacrylate) (P2,6DMFM), Poly(2,4-dimethylphenyl methacrylate) (P2,4DMFM), Poly(2,5-dimethylphenyl methacrylate) (P2,5DMFM), Poly(3,5-dimethylphenyl methacrylate) (P3, 5DMFM). (From ref. [42])

and gives reproducible results in terms of polymer transitions. The low equivalent
t.s.c. frequency ($\sim 10^{-4}$ to 10^{-2} Hz) leads to enhanced resolutions of the different
relaxation processes, especially the β relaxation process. A thermal peak cleaning
of the t.s.c. global spectra is often used to study broad relaxation peaks as the low
temperature secondary relaxation peaks.

There are at least two ways to obtain information by t.s.c. The first is that for the
global spectra where the polymeric film is polarized by a static electric field at the
polarization temperature Tp and then quenching down to the freezing temperature.
With the field turned off and the sample short circuited the depolarization current
due to dipolar reorientation is measured as the temperature increases from T_0 to the
final temperature $T_f \geq T_p$.

The second way is to apply the polarization field during the following thermal
cycles.: maintain the sample at T_p, quenching to $T_p - 10$ ang for a while. The
field is the removed and the sample is quenched to a temperature $T_0 = T_p - 40$.
The depolarization spectra due to a narrow distribution of relaxations the are then
measured upon heating a certain rate above T_p [42].

The global t.s.c. is a convoluted spectrum of all dielectrically active relaxations
excited between T_p and T_0. The relaxation time $\tau(T)$ is related to the measured
depolarization current iT by:

$$\tau(T)\frac{P(T)}{i(T)} = \frac{\int_{T_0}^{T} i(T)d}{i(T)t} \qquad (2.40)$$

Where T_0 is the initial temperature of the depolarization scan. It is assumed that
the relaxation time constant τ is related to the barrier height or apparent activation
energy E_a in the Arrhenius equation.

The second method of analyzing the thermally cleaned spectrum uses the Eyring
equation:

$$f = \frac{kT}{2\pi h}\exp\left(-\Delta H^{\neq}/RT\right)\exp\left(\Delta S^{\neq}/RT\right) \qquad (2.41)$$

where k is Boltzmann`s constant, h is Plankcs's constant and $\Delta H^{\neq}$ and $\Delta S^{\neq}$ are the
Eyring activated states of enthalpy and entropy, respectively [42].

The imaginary part of the dielectric constant ε'' at 1 Hz is plotted against T in
Figs. 2.49 and 2.50 shows the global t.s.c.multiplot for PDMPM. The t.s.c depolarization current is analogous to a conventional dielectric loss signal with an equivalent frequency of the order of 10^{-3} Hz. Therefore, the shape of each spectrum is
similar, except for the very high temperature peaks in t.s.c. thermograms which do
not appear in dielectric a.c. spectra. They are due to free charges and are called ρ
peaks.

The low temperature zone correspond to the β relaxation; the absorption is weak
and is extended over a large span of temperatures. The β process is suggested to be
associated with the motions of the entire side chain groups [156–158].

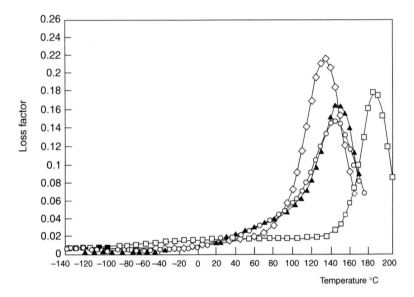

Fig. 2.49 Dielectric loss factor at 1 Hz: (□) 2,6-PDMPM; (○) 3,5-PDMPM; (◇) 2,5-PDMP; (▲) 2,4 PDMPM. (From ref. [41])

The spectra show prominent absorption at high temperature with high and sharp peaks attributed to the glass-rubber transition. The last peak in t.s.c. spectra correspond to the conductivity of the material; these are the so called ρ peaks and are present at high temperature.

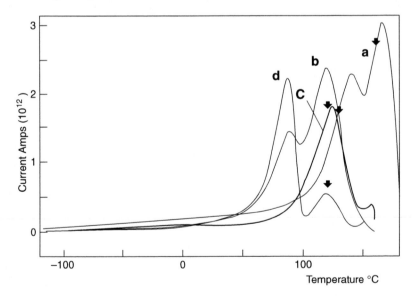

Fig. 2.50 Full t.s.c spectra the four polymers i.e. (**a**) 2,6-PDMPM, at 167°C; (**b**) 2,4-PDMPM at 109°C (**c**) 3,5-PDMPM at 106°C; (**d**) 2,5-PDMPM at 111°C. (From ref. [41])

It is interesting to note that 2,6 substitution on the aromatic ring leads to very high steric hindrance [42, 158]. The 2,6-PDMP α -temperature zone is detected higher by about 50°C when compared with the other polymers of this group. Differences of only 3 or 4°C in the temperature zone are observed among 2,4-PDMP, 3,5-PDMP and 2,5-PDMP. The main characteristic of 2,5-PDMP is that the dipoles introduced with the methyl substituent cancel each other because they diverge in direction bt 180°, consequently interaction with the -COO group dipole is absent.

The α zone of this polymer lies at the lower temperatures, what is attributed to the lack of dipole-dipole interaction what make easier the segmented mobility. Concerning the β relaxation, 3,5-PDMP present the most important absorption and 2,5-PDMP have the weakest process in the glass system.

Complex dielectric plane plots for 2,6-PDMP, 2,4-PDMP, 3,5-PDMP, 2,5-PDMP, are shown in Fig. 2.51. (taken from ref. [42]). In this case the experimental arc is obtained by shifting the isothermal curves of ε' and ε'' in frequency by application of the time-temperature superposition principle, effecting a horizontal and eventually a vertical shift of each one. This is made possible by the relaxed and unrelaxed permittivities (ε_0 and ε_∞) being nearly temperature independent. In all cases, the α absorption is represented by skewed arcs which approach the abcissa in the high frequency region through a straight line. As in the previous systems curves have been fitted using the Havriliak-Negami [70] empirical equation (2.30) and by the biparabolic equation [159–162].

$$\varepsilon^*(w) = \frac{\varepsilon_0 \varepsilon_\infty \left[1 + \delta \, (jw\tau)^{-k} + (jw\tau)^{-h} \right]}{\varepsilon_0 - \varepsilon_\infty + \varepsilon_\infty \left[1 + \delta \, (jw\tau)^{-k} + (jw\tau)^{-h} \right]} \qquad (2.42)$$

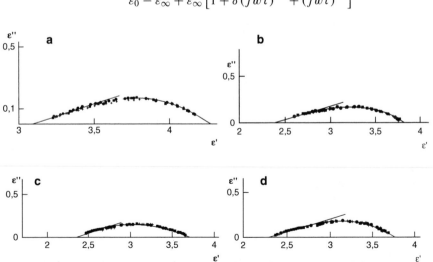

Fig. 2.51 Cole-Cole plots for the polymers studied (+) experimental data; (■) Havriliak-Negami model; (●) biparabolic model. References temperatures: (**a**) 2,6-PDMPM, 167°C; (**b**) 2,4-PDMPM, 109°C; (**c**) 3,5-PDMPM, 106°C; (**d**) 2,5-PDMPM, 111°C. (From ref. [42])

The mechanical counterpart of this equation comes from a theory of the non-elastic deformations developed from a molecular mobility model. The parameters δ, h and k have precise physical meanings, taking into account the effectiveness of correlation effects exhibited during the molecular motions involved in the process.

From the comparison of the three arcs, shown in Fig. 2.51, the biparabolic equation fits the experimental data better than the Havriliak – Negami equation (2.28), even though the two models give good results. The polymer 2,6-PDMP has the weakest intensity ($\varepsilon_0 - \varepsilon_\infty$), meaning that its process of α relaxation involve fewer dipoles than the others. This is consistent with the fact that the glass transition is higher at higher temperature. The molecular chains must pass through high potential barriers, so they have many difficulties to overcome in reorienting their dipoles. 2,5-PDMP has the highest value of $\varepsilon_0 - \varepsilon_\infty$ and the smallest α zone of temperatures.

Obviously in the limit of low and high frequencies the parameters of the two models determine the frequency dependence of the dielectric loss in the α relaxation [162, 163].

Therefore they are related by a mathematical equation: at very high frequencies ($w \rightarrow \infty$)$k = \beta(1 - \alpha)$ and at very low frequencies ($w \rightarrow 0$), $k = (1 - \alpha)$.

k and $\beta(1 - \alpha)$ values depend on the polymer behavior in the range of high frequencies for the α relaxation i.e. inducing the easiest and fastest coopèrative motions. On the other hand, h and $1 - \alpha$ depend on slower movements, higher temperatures, lower frequencies. Therefore, the quantitative values of these parameters give information about these emotions in proportion to the global α process.

As in the cases previously mentioned, the α processes are assigned to the cooperative motions around Tg. Because of these motions the α process can follow a WLF behavior for the shift factor α_T of the isothermal curves ε''.

The values for ϕ_g, the free volume at the glass transition temperature, are slightly different from the predicted theoretical value (0.0252 ± 0.005). 2,6-PDMP is the polymer with the highest value of ϕ_g, and this value decreases with temperature owing to reduced molecular motions. Since the chains of 2,6-PDMP are the slowest, the volume set at the glass transition temperature should be the highest .The value for 2,5-PDMPis quite high especially when is compared with the value of 3,5.-PDMP. The chains of the former are faster and the free volume of the latter is larger as a result of the two substitutions at positions 3 and 5. The α' -relaxation process is attributed to the micro-Brownian motions of the main chain, involving long cooperative motions. These cooperative motions are a succession of correlative movements in the whole structure which depends on the configuration that the chains take at each instant and on the available volume.

The main-chain mobility depends slightly on the ease of the lateral group motion, so the respective positions of the methyl groups influence this mobility because of their important steric volume [157].

Stronger steric hindrance results from ortho substitution : the phenyl groups are highly restricted to rotation, and therefore the structure becomes inflexible and the chains cannot change configuration easily. This is the case for 2,6-PDMP with the methylgroups in both available ortho positions [42].

The β zone extend over a large temperature range. This is a characteristic of a secondary process which involve local motions of the lateral groups [155]. They are more diversified movements with a large spectrum of relaxation times. Therefore, thermal cleaning of the t.s.c. global spectra is used to study the broad relaxation peaks of the low temperature secondary relaxation [42]. This is effective because it allows one to excite only the specific transition of interest [155].

Figure 2.52 shows the thermally cleaned spectrum for 2,6-PDMP. Each peak corresponds to a window of polarization of 10°C and to an extremely narrow

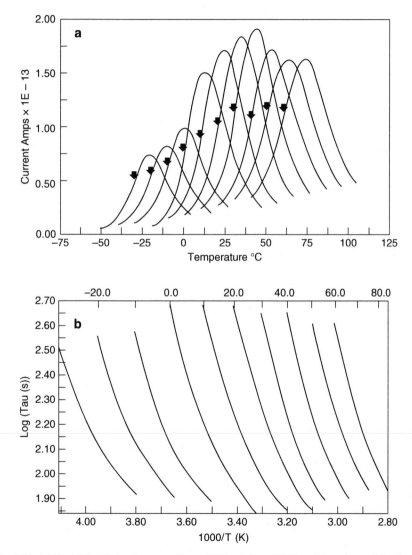

Fig. 2.52 (**a**) Partial depolarization according to Tables 2.11 and 2.12 and experimental for details (**b**) relaxation maps analysis of 2,6-PDMP. (From ref. [41])

Table 2.11 Arrhenius parameters for the polymers studied at the polarization temperature T_p and the temperature at the maximum of each peak T_m. (From ref. [42])

Polymer	T_p (°C)	T_m (°C)	$\log \tau$	E_a (kcal mol^{-1})
2,6-PDMP	−30.0	−20.5	−5.052	8.69
	−20.0	−9.6	−5.398	9.48
	−10.0	0.8	−5.605	10.13
	0.0	12.6	−5.976	11.11
	10.0	14.0	−6.222	11.88
	20.0	34.4	−6.341	12.46
	30.0	43.4	−6.868	13.61
	40.0	53.1	−6.821	13.99
	50.0	62.7	−6.472	13.85
	60.0	73.1	−6.859	14.90
2,4-PDMP	−40.0	−29.0	−6.084	9.48
	−30.0	−19.7	−6.532	10.39
	−20.0	−9.5	−6.532	10.82
	−10.0	0.6	−7.661	12.67
	0.0	11.8	−7.772	13.37
	10.0	23.3	−7.234	13.18
	20.0	33.6	−9.056	16.20
3,5-PDMP	−40.0	−26.4	−6.645	10.25
	−30.0	−17.2	−6.634	10.66
	−20.0	−6.3	−7.017	11.59
	−10.0	2.9	−6.578	11.42
	10.0	23.6	−6.573	12.34
	20.0	33.5	−6.916	13.27
2,5-PDMP	−40.0	−27.8	−7.142	10.73
	−30.0	−19.9	−8.600	12.74
	−20.0	−9.8	−7.236	11.66
	−10.0	0.1	−8.133	13.25
	0.0	9.7	−7.818	13.30
	10.0	20.8	−8.605	14.95

relaxation time distribution. The plot of $\ln \tau$ against $1/T$ resulting from each peak integration allow one to determine the activation parameters of the process. Values of these parameters are shown in Tables 2.11 and 2.12 for the polymers studied.

E_a is related to $\Delta H^{\neq}$ by $E_a = \Delta H^{\neq} + RT$. The value of RT is of the order of $0.5 \, \text{kcal mol}^{-1}$ at normal temperatures and is only a small contribution. That is why the values of E_a and $\Delta H^{\neq}$ are so close. These two parameters represent the energy barrier that the chain must pass through during their movements. The values are in agreement with a secondary relaxation but too high for only local and independent motions of the local groups [122].

The activation entropy correspond to the density fluctuations of the material during the relaxation process [122, 163, 165].

The initial hypothesis is that the β -relaxation process of the PDMP involved local and independent motions of the side chains, taking into account the broadness

Table 2.12 Thermodynamic parameters. (From ref. [42])

Polymer	T_p (°C)	T_m (°C)	ΔH (kcal mol^{-1})	ΔS (cal deg^{-1} mol^{-1})	ΔG (kcal mol^{-1})
2,6-PDMP	−30.0	−20.5	7.8545	−37.0033	16.8463
	−20.0	−9.6	8.5041	−35.5008	17.5858
	−10.0	0.8	9.1987	−34.6278	18.3059
	0.0	12.6	10.1249	−33.0353	19.1435
	10.0	24.0	10.8413	−31.9776	19.8910
	20.0	34.4	11.3825	−31.5165	20.6168
	30.0	43.4	12.4747	−29.1250	21.2998
	40.0	53.1	12.8066	−29.4276	22.0175
	50.0	62.7	12.6668	−31.0938	22.7101
	50.0	73.1	13.6452	−29.3783	23.4282
2,4-PDMP	−40.0	−29.0	8.6521	−32.2871	18.1750
	−30.0	−19.7	9.5056	−30.2150	16.8479
	−20.0	−9.5	9.9120	−30.2856	17.5743
	−10.0	0.6	11.6532	−25.1976	18.2801
	0.0	11.8	12.3045	−24.7909	19.0724
	10.0	23.3	12.0812	−27.3532	19.8221
	20.0	33.6	14.9727	−19.0848	20.5646
3,5-PDMP	−40.0	−26.4	9.3836	−29.6625	16.2949
	−30.0	−17.2	9.7383	−29.8055	16.9810
	−20.0	−6.3	10.6185	−28.1440	17.7391
	−10.0	2.9	10.4520	−30.2201	18.3998
	10.0	23.6	11.2616	−30.4200	19.8706
	20.0	33.5	12.1385	−28.9303	20.6151
2,5-PDMP	−40.0	−27.8	9.8439	−27.4097	16.2303
	−30.0	−19.9	11.7423	−20.8484	18.8079
	−20.0	−9.8	10.7118	−27.0967	17.5672
	−10.0	0.1	12.2191	−24.5973	18.9446
	0.0	9.7	12.2295	−24.5973	18.9446
	10.0	20.8	13.8095	−21.0038	19.7536

of the spectra and the distributions of the values for the activation parameters. The motions are much more complicated that intermolecular and intramolecular motions occur. A change in conformation or lateral rotation does not occur without a correlated motion of that part of the main chain close to the side group or without a change of conformation of a neighbouring lateral group. These intermolecular hindrances to reorientations contribute to the energy barriers and lengthen of the relaxation time. The two methyl groups have an influence on the flexibility of the lateral groups and on the free volume of the glassy structure.

In the case of 3,5 substitution, the excluded volume from the rotation of the disubstituted phenyl is large leading to an increased free volume. The local motions thus becomes easier in the glassy state. On the contrary, the 2,5 substitution, for which the net effect of the steric hindrance, the overall bulkiness of the substituents and absence of dipole-dipole interaction result in a higher segmental mobility, has not an important available free volume in the glassy structure. The first local movements cannot begin easily at low temperature and the β relaxation is nonexitent.

The secondary process depends on the broadness of the motions and the nonequilibrium of the polymer during the polarization. Indeed 2,6 PDMP, whose molecules are the lowest, present a weak β process. The relaxation involve few dipoles with small motions. Moreover, the chains are closer to their equilibrium state than the chains of the other polymers and the energy barrier is lower because the activation parameters are lower.

The molecular mechanism of this secondary relaxation is not well understood but the important source is thought to be reorientation associated with flexible side chains [153, 162, 163]. The movements involved at low temperatures are local but depend on the variation of the packing in the glass about individual reorientating groups. Therefore, there is a cooperativity in the movements and the correlation increases with temperature [163, 164].

(b) Dynamic mechanical relaxational behavior
As it has been described previously, dynamic mechanical behavior and viscoelastic relaxation in polymers are closely related to their chemical structure. It is very well known that poly(phenylmethacrylate) (PPHM) does not show significative viscoelastic activity below Tg. On the other hand the presence of –CH$_2$ spacer groups as in cases previously mentioned, for saturated side chain rings containing polymers, promotes a larger molecular mobility giving rise at least at two viscoelastic relaxion, like in poly(benzyl methacrylate)s (PBzM) but with low intensity [29]. Likewise in these systems the glass transition temperature diminishes about 50°C, relative to that of poly(phenyl methacrylate) (PPHM) [29]. Dielectric studies relative to these systems show that beside [30, 42] the viscoelastic responses of these poly(methacrylate)s [42] present two slightly dielectric absorptions below Tg. Besides the above considerations relative to secondary relaxations, it is also important to analyze the influence of the substituents in the side group on the glass transition temperature. As it was mentioned above, when the flexibility of the side chain increases, the Tg values diminishes. Therefore, the effect of the side chain structure on the viscoelastic bevahior of these aromatic polymers is an interesting aspect to be taken into account, particularly in the case of isomer polymeric materials, i.e. polymers containing aromatic side chains with substitutions with the same group but in different positions of the aromatic ring. Poly(dimethyl phenyl methecrylate)s is an example of these kind of chemical structures.

Figure 2.53 show the real part of the dynamic modulus and loss tangent at 1 Hz for poly(2,6-dimethylphenyl methacrylate) (2,6PDMP), poly(2,4-dimethylphenyl methacrylate) (2,4PDMP), poly(2,5-dimethylphenyl methacrylate) (2,5PDMP) and poly(3,5 dimethylphenyl methacrylate) (3,5PDMP). Its is possible to observe that 2,6PDMP is the polymer which show the higher Tg value and 2,5PDMP the lower what can be summarized in the next Table 2.13. (From ref. [41]).

Another fact to take into account is that there is no dipole-dipole interaction between both methylene groups due to the antiparalell position of them. For this reason it is possible to think that the glass transition temperature of 2,5 PDMP should be

Fig. 2.53 Dependence of the storage modulus and loss tangent with temperature for 2,4PDMP (■), 2,5-PDMP(▼), 2,6 PDMP(●) and 3,5-PDMO(△). (From ref. [41])

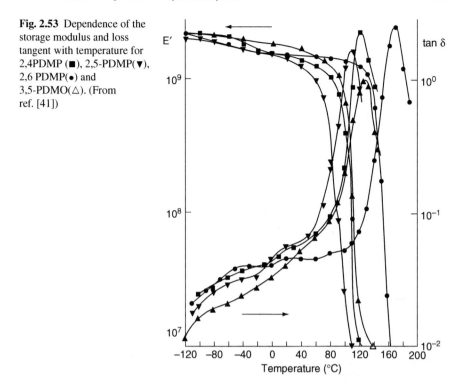

lower because of the higher mobility [41]. These arguments are in good agreement with the results shown in Table 2.14. In the vitreous zone there is some viscoelastic activity around $-40°C$ and $20°C$.

It is noteworthy the comparison with monosubstituted poly(phenyl acrylate) (PPhA) containing chlorine in the aromatic ring, it has been observed [154] intensive subglassmechanical absorption which have been attributed to the whole rotation of the side group. As the side of the chlorine group is similar to that of the methyl group, it should be expected the same relaxations in the corresponding monmethyl substituted poly(phenyl acrylate)s and also poly(dimethylphenyl) derivatives. On the contrary in the case of poly(alkylphenyl methacrylate)s the absence of that absorption should be attributed to the restriction to molecular mobility due to the α methyl group in poly(methacrylate) chains. In fact, it is very well known [28, 152]

Table 2.13 Dependence of the temperature of the maximum in $\tan \delta$ at 1 Hz for the α relaxation as function of the position of the methyl substituents for 2,4 PDMP, 2,5 PDMP, 2,6 PDMP and 3,5PDMP. (From ref. [41])

Polymer	$T/°C$
2,4 pdmp	121
2,5 PDMP	108
2,6 PDMP	167
3,5 PDMP	127

Table 2.14 Summary of the dielectric results for poly(2,4-difluorobenzyl methacrylate) (P24DFBM), poly(2,5-difluorobenzyl methacrylate) (P25DFBM), and poly(2,6-difluorobenzyl methacrylate) (P26DFBM). (From ref. [31])

Polymer	$d\varepsilon/dw$	$2n\,dn/dw$	$\langle\mu^2\rangle/x$, D^2
P24DFBM	1.38	0.09	2.50
P25DFBM	1.09	0.12	1.92
P26DFBM	2.65	0.10	5.00

that the α methyl groups in these systems not only increases considerably the glass transition temperature, but they also reduce the mobility of the groups which would be the responsible of the relaxation associated to the Tg.

According to the results dealing with poly(benzyl methacrylate)s [29] it is clear that the absence of methyl substituents on the phenyl ring and the presence of a –CH$_2$ spacer group increases the molecular mobility, showing two secondary relaxation. Nevertheless they are not enough to compensate the large steric hindrance due to the α -methyl group.

2.4.4.2 Poly(difluorobenzylyl methacrylate)s

There exist enormous interest in the study of fluorinated polymers with the fluorine atoms either attached to the main chain or located in the side chains, mainly from industrial point of view [166]. The cause of this interest lies in the resistance of most of these polymers to acids and alkalis and their insolubility in common organic solvents [167]. Moreover, fluorine atoms provide excellent dielectric properties and weatherability, low coefficient of friction, and chemical inertness to polymers. Because of the high polarity of the fluorine atoms, small modifications on the structure arising from the location of these atoms in the chains give rise to important differences in the dielectric relaxational behavior of fluorinated polymers [31]. These polymers are of interest due to their inherent properties such as hydrophobicity, low surface tension and so forth, and also for the unique structure frequently generated and the properties arising from the ordering of the side chains [31, 168–170]. As a consequence of the useful physical properties of fluorinated polymers, material of this kind are being used for industrial applications [31, 160].

The relaxational behavior of fluorinated polymer is an interesting example for the analysis of subglass relaxation characterized by an Arrhenius type temperature dependence of the relaxation times and broad distribution of relaxation times [171, 172]. The relaxation response is dependent on the experimental probe, and for a critical interpretation of the dynamics of molecular chains it is necessary in addition to the use of different problems, experimental data on polymer chains in which the response is sensitive to conformational changes of determined bonds. This analysis was performed by Díaz-Calleja and coworkers [31] using difluorinated phenyl isomers of poly(benzyl methacrylate)s (PBM), specifically, poly(2,4-difluorobenzyl methacrylate) (P24DFBM), poly(2,5-difluorobenzyl methacrylate) (P25DFBM) and poly(26PDFBM). A sketch of the repeating unit of P24DFBM is shown in Scheme 2.12.

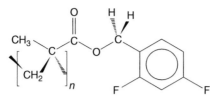

Scheme 2.12 Sketch of the repeating unit of poly(2,4-difluorobenzyl methacrylate) (P24DFBM). (From ref. [31])

Due to the high polarity of these polymers the location of the fluorine atoms in the aromatic ring play an important role on the molecular motions below glass-rubber transition. For this reason the knowledge of the mean square dipole moment per repeating unit, $\langle \mu^2 \rangle / x$, which is calculated by means of the Guggenheim- Smith equation [173–175]:

$$\frac{\langle \mu^2 \rangle}{x} = \frac{27 k_B M_0 T}{2\pi \rho N_A (\varepsilon_1 - 2)^2} \left(\frac{d\varepsilon}{dw} - 2n \frac{dn}{dw} \right) \tag{2.43}$$

where k_B and N_A are, respectively, the Boltzman constant and Avogadro's number, M_0 is the molecular weight of the repeating unit, and ρ and ε_1 are, respectively, the density and the dielectric permittivity of the solvent.

Values of $d\varepsilon/dw$ and $2n \, dn/dw$ for the fluorinated polymers are shown in Table 2.14

It is interesting to note that the term $d\varepsilon/dw$ for P26DFBM, which is proportional to the total polarization of the chains, is nearly two times the values of this quantity for the other polymers. As expected, the term $2n_1 dn/dw$, proportional to the electronic polarization, is in comparison with $d\varepsilon/dw$ only slightly dependent on the chemical structure. The polarity of the polymers, expressed in terms of the mean-square dipole moment per repeating unit, is very sensitive to the location of the fluorine atoms in the aromatic ring, as it was mentioned above. This can be seen also in Table 2.14 according to the values of $\langle \mu^2 \rangle / x$.

Dielectric relaxational behavior

The variation of the dielectric loss with temperature at different frequencies for P26DFBM is shown in Fig. 2.54 at low frequencies, the isochrones exhibit a well defined subglass γ absorption followed in order of high temperature for the β process that overlaps with the low temperature side of the glass-rubber relaxation. The γ relaxation is poorly defined for P24DFBM as can be seen in Fig. 2.55 where the values of ε'' at 100 Hz, are plotted against temperature for the three fluorinated polymers. The isochrones present a prominent glass-rubber, or α relaxation, that in the case of P26DFBM is masked by conductivity contributions at frequencies below 10 Hz. This behavior is also observed in other systems previously reported. The intensity of the α relaxation of P26DFBM, measured in terms of the height of the peak, is about three times that of the glass-rubber relaxation of P24DFBM

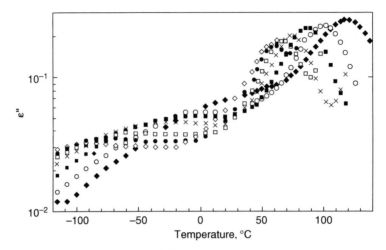

Fig. 2.54 Experimental loss perimttivity (ε'') as a function of temperature for P24DFBM at several frequencies: ($\blacklozenge$) 10^5, ($\bigcirc$) 10^4, ($\blacksquare$) 10^3, (X) 10^2, ($\square$) 10^1, ($\bullet$) 10^0, and ($\Diamond$)10^{-1} Hz. (From ref. [31])

and P25DFBM. The relaxation maps corresponding to these polymers are shown in Fig. 2.56.

The glass-rubber relaxation are located in the mechanical spectra, at 1 Hz in the vicinity of the respective calorimetric glass-transition temperatures. The intensity of the α peak for P24DFBM, expressed in terms of the height of the loss peak, is slightly higher than the intensities of the peaks corresponding to P25DFBM and P26DFBM.

The molecular origin in polymers with flexible polar groups is generally attributed to motions of these groups [176]. The fact that the γ relaxation is poorly defined in P25DFBM where the two C^{ar}—F dipoles of the phenyl groups cancel out suggests that this relaxation in P24DFBM and P26DFBM is associated to the motions of the CH_2–C^{ar} bonds in the side groups. The overlapping of the β and α relaxations are indicative that motions in the whole side groups may in-

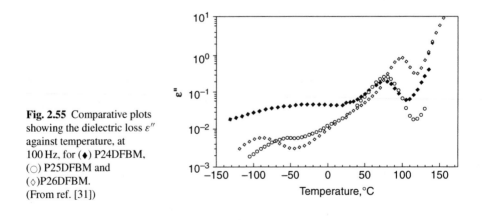

Fig. 2.55 Comparative plots showing the dielectric loss ε'' against temperature, at 100 Hz, for ($\blacklozenge$) P24DFBM, ($\bigcirc$) P25DFBM and ($\Diamond$)P26DFBM. (From ref. [31])

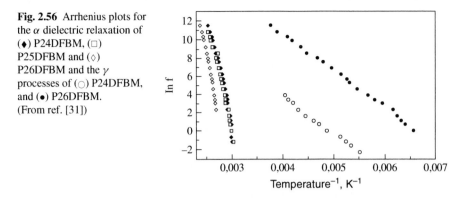

Fig. 2.56 Arrhenius plots for the α dielectric relaxation of (♦) P24DFBM, (□) P25DFBM and (◊) P26DFBM and the γ processes of (○) P24DFBM, and (●) P26DFBM. (From ref. [31])

tervene in the development of the former process. These motions coupled with the micro-Brownian motions of the main chain, surely also intervene in the α relaxations.

The fact that the three polymers behave on a rather different way despite their structural similarities may be explained by analyzing the conformational and dipolar properties of the model compound of the repeating units. Molecules like 2,4-, 2,5-, and 2,6-difluorobenzyl 2,2-dimethyl propionates (2,4, 2,5 and 2,6 DFP) have been employed for the conformational analysis [31]. Scheme 2.13 Is a representation of the model molecules.

Charges assigned to every atom are employed both to compute the Coulombic term of conformational energies and the molecular dipole moment. Nevertheless, arrows in Scheme 2.13 represent main contributions to the molecular dipole moment arising from the ester group and the C^{ar}—F bonds which are used only for qualitative analysis [31] The molecular calculation strategy is performed following common procedures for this purpose and the appropriate computational packages [31, 177–179]. A great amount of conformations for each molecule is obtained by rotation over the CO—CH_2C^{ar} (ϕ_1) and OCH_2—$C^{ar}C^{ar}$ (ϕ_2) bonds.

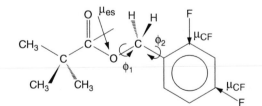

Scheme 2.13 Rough sketch of the molecule 2,4-difluorobenzyl 2,2-dimethylpropionate (2,4-DFP) used as model compound for the repeating unit of the corresponding polymer. Rotations over CO—CH_2C^{ar} (ϕ_1) and OCH_2—C^{ar} C^{ar} (ϕ_2) bonds are drawn in their planar trans conformation which is represented by $\phi_1 = \phi_2 = 180°$. *Arrows* represent main contributions to the molecular dipole moment, arising from the ester group and the C^{ar}—F bonds. These contributions are indicated only for illustrative purposes, since actual calculations of dipole moments were performed with partial charges assigned to every atom of the molecule. (From ref. [31])

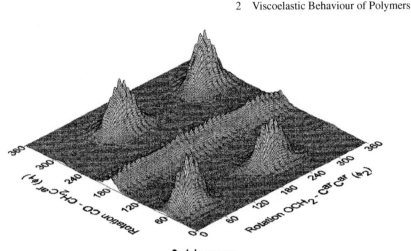

2,4 Isomer

Fig. 2.57 The distribution of probabilities, computed at 25°C for the conformations of 2,4-DFP obtained by rotation of ϕ_1 and ϕ_2 angles. (From ref. [31])

The results of this procedure indicate that the three molecules are quite similar as far as the conformational freedom is concerned. For instance, conformational partition functions evaluated by Diaz Calleja et al. [31] amount to 68.3, 70.6 and 61.5, respectively, for 2,4-, 2,5-, and 2,6-DFP molecules at 25°C. The distribution of probabilities, computed at 25°C for the conformations of 2,4-DFP obtained by rotation of ϕ_1 and ϕ_2 angles are represented in Fig. 2.57.

The results give four areas of high probability plus a crest obtained with any value of ϕ_2 provided that $\phi_1 \approx 180°$ indicating that the phenyl ring is almost freely rotating when the whole ester group is in the planar trans conformation. On the contrary, the probability becomes rather low, i.e. the energy is high, when $\phi_1 \approx 0_1$ regardless of the value of ϕ_2 or, in other words the cis conformation of the CO—CH$_2$C^{ar} bond are forbidden. Although the conformational characteristics of these three molecules are quite alike, their dipole moments are rather different, as can be seen in Table 2.15.

Therefore the comparison of the dielectric spectra of P25DFBM with those of P24DFBM and P26DFBM leads to conclude that the motions about the C^{ar}—CH$_2$ bonds are involved in the development of the subglass γ relaxation. The overlapping of the α and the β relaxations that partially or totally mask the latter process

Table 2.15 Summary of the dipolar results obtained for 2,4-, 2,5-, and 2,6-DFP Molecules. (From ref. [31])

Molecule	μ_{min}, D	μ_{max}, D	$\langle \mu^2 \rangle$, D^2	$\langle \mu^2 \rangle^{1/2}$, D
2,4-DFP	0.31	3.66	6.05	2.46
2,5-DFP	1.29	3.09	3.35	1.83
2,6-DFP	0.35	4.47	4.40	2.10

Averaged values were computed at 25°C.

suggests that motions occurring in the whole side chain presumably intervene in the β relaxation.

The dielectric relaxation strength of the polymers follows the same trends as the mean-square dipole moments per repeating unit. Thus P26DFBM and P25DFBM which have the higher and lower values of $\langle \mu^2 \rangle / x$, 5.00 and 1.92 D^2 at 25°C, respectively.

The fact that the glass-transition temperature of the high polarity P26DFBM and P25DFBM, in spite of the similarity of the flexibility of the three polymers, hints to dipolar intermolecular intermolecular interactions as responsible for the differences observed between the Tg's of the phenyl difluorinated. On the other hand Molecular Mechanics give a good account of the differences observed in the polarity of the three polymers containing fluorinated phenyl moieties assuming that the side groups are in planar or nearly planar conformation.

2.4.4.3 Poly(dichlorobenzyl methacrylate)s

Another interesting family is that of poly(monochlorobenzylmethacrylate)s (PM-ClBM and poly(dichlorobenzylmethacrylate)s (PDClBM). Scheme 2.14 Show the chemical structures of these polymers. The dielectric relaxational behavior of these polymers, studied by determining the components of the complex dielectric permittivity ε^*. Two relaxation processes, labelled as α and β relaxations, can be observed in Fig. 2.58. Both relaxations are associated, as in all the previous systems, to the glass transition temperature α and β as a shoulder of the main relaxation. As in previous systems, at low frequencies and high temperatures a conductive contribution can be observed. To split α and β relaxations the dielectric loss ε'' is described using the Hopping model [39, 86]:

$$\varepsilon'' = \frac{\sigma}{\varepsilon \omega^S} \tag{2.44}$$

where σ is the conductivity, ε is dielectric permittivity of the vacuum, ω is the angular frequency and s is a parameter to be determined. In this case ionic conductivity predominates over partial blocking phenomena.

The activation energy as in previous systems is analyzed using an Arrhenius type plot like that represented in Fig. 2.59 [39]. Following the strategy reported by Díaz Calleja et al. [39] it can be concluded β that relaxation behavior of PMCBM and PDCBM is complex and in general these polymers show two main relaxations associated to the dynamic glass transition temperature and a diffuse relaxation. According to the results reported in ref. [39], the relative position of the chlorine atoms in the benzyl ring influences directly the position and the strength of the secondary relaxation. This results is indicative that the benzyl ring take part on the β relaxation and the relative position of the chlorine atoms determines the glass transition temperature. The mechanical measurements shown in Fig. 2.60, of PDCBM confirm these results [39].

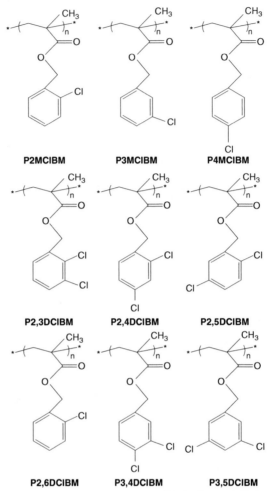

Scheme 2.14 Chemical structures of PMCBM and PDCBM. (From ref. [39])

The mean-square dipole moment depends on the relative position of the chlorine atom. The mean-square dipole moment for P3MCBM and P2,4DCBM are similar, and the same is true for P4MCBM and P3,5DCBM. As the dipole moment is related to the relaxation strength by means of the Onsager-Fröhlic-Kirkwood equation (OFK) [181, 182] the α and β relaxations using this equation show a rough estimation of the of the combination factor for the combined α and β relaxations. The results show that the correlation factors with those obtained in solution for polymers with similar structure, suggests that the intramolecular correlations are dominant in these systems.

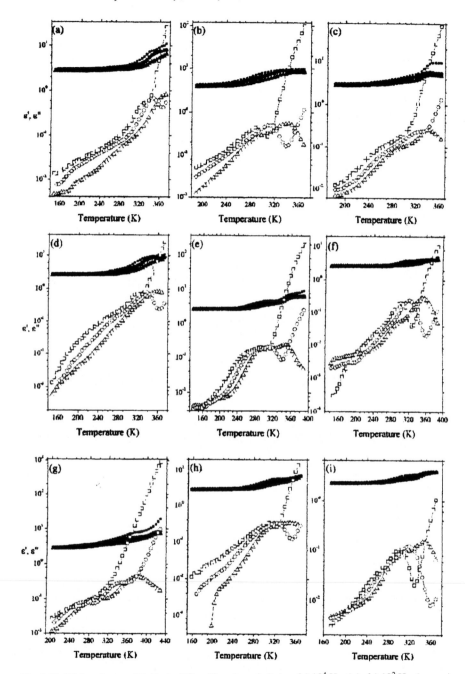

Fig. 2.58 Dielectric permittivity (*solid*) and loss (*open*), (*triangle*) 10^4 Hz, (*circle*) 10^2 Hz, (*square*) 10^0 Hz. (**a**) P2MCBM, (**b**) P3MCBM, (**c**) P4MCBM, (**d**) P2,3 DCBM, (**e**) P2,4DCBM, (**f**) P2,5DCBM, (**g**) P2,6DCBM, (**h**) P3,4DCBM, and (**i**) P3,5DCBM. (From ref. [39])

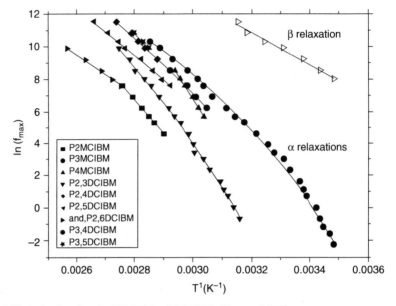

Fig. 2.59 Arrhenius plots for PMCBM and PDCBM. (From ref. [39])

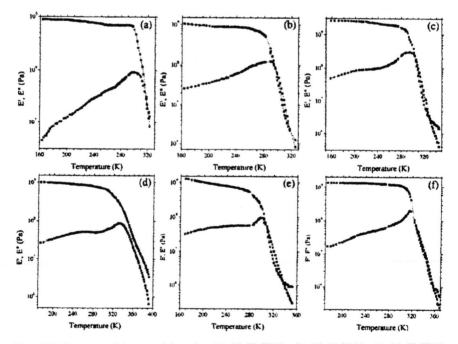

Fig. 2.60 Storage and loss moduluss for (**a**) P2,3DCBM, (**b**) P2,4DCBM, (**c**) P2,5DCBM, (**d**) P2,6DCBM. (From ref. [39])

2.5 Viscoelastic Properties of Poly(itaconate)s

Itaconic acid is a bifunctional monomer obtained by the fermentation process using *Aspergilius itaconicus and Aspergilius terreus fungi* [182, 183].

Esterification of itaconic acid is a very interesting way to obtain monomers that present an attractive array of possible structural variations. Mono and diesterification of itaconic acid can be carried out, obtaining monomers and polymers having either one or two of the carboxyl groups esterified in each monomer. Monoesters can also be selectively esterified in order to obtain diesters with two different side groups, specifically methyl-alkyl-diesters can be obtained [182, 183]. Mono and diesterification of itaconic acid can be carried out, obtaining monomers and polymers having either one or two of the carboxyl group esterified in each monomer. Monoesters can also be selective esterified in order to obtain diester with two different side groups, specifically methyl-alkyl diesters [184].

The variation of the nature of the ester group, depending on mono and diesters derivatives provides several different polymer series with very interesting properties [183–196].

Polymerization of these monomers is commonly achieved by radical polymerization in solution and/or in bulk. Poly(monoitaconate)s may be considered as typical comblike-polymers depending on the length of the side chain in the case of aliphatic derivatives [191, 192].

The effect of the length of the side chain and the presence of the carboxylic group have been taken into account to explain the particular conformational behavior of some of these poly(itaconate)s [191, 192]. The effect of the length of the side group and the presence of the carboxylic group are the responsible of the particular conformational behavior of some poly(monoitaconate)s [191, 192]. On the other hand poly(di-n-alkyl itaconate)s obtained from diester of itaconic acid and lower unbranched alcohols, show important changes in both solution and the solid state properties [182–185].

The chemical versatility of this family of polymers represent an interesting field to analyze the dynamic-mechanical and dielectric analysis of polymers. The fact that is possible to obtain polymers with mono and disubstitution is a good way to be able to know which is the effect of the variation of the fine chemical structure on the viscoelastic responses of these materials. In fact, mono and disubstituted structures should be interesting in order to analyze the effect of the side chain structure on the relaxational behavior of these systems, and of course, to get confidence about the origin of the molecular motions responsible of the conformational changes that take place in these systems. By this way it is possible to obtain important information about the relaxation processes that can be observed in substituted vinyl polymers. The thermal stability, dielectric, dilute solution behavior, poly(electrolyte)s and ion exchange behavior, biological and ati-tumor behavior and response to high –energy radiation, viscoelastic properties of members of several series of poly(itaconate)s and copolymers containing itaconate units has been assessed by several authors [197–200, 200–237].

2.5.1 Poly(itaconate)s Containing Aliphatic and Substituted Aliphatic Side Chains

The viscoelastic behavior of poly(mono-n-octylitaconate) (PMOI) and poly(mono-β n-decyl itaconate) (PMDI) (see Scheme 2.15) shows three relaxations peaks between -120 and $140°C$ and a prominent subglass relaxation is observed near $-60°C$. In this system a slight viscoleastic activity labeled as β' is observed at room temperature and then a high temperature peak (α) probably related to the glass transition temperature as can be observed in Figs. 2.61 and 2.62. This is a behavior similar to that observed for amorphous polymers [237]. A similar viscoelastic response is observed in both polymers. A weak relaxation phenomenon is present near room temperature. The activation energies can be obtained using the typical Arrhenius plot of Ln f_m vs T^{-1} which in both systems are $25\,kcal\,mol^{-1}$. These values are obtained from the plots shown in Fig. 2.63.

In these systems, it is interesting to note that they show two new relaxations at higher temperatures than those for the former dielectric measurements in PMOI and PMDI. One of them can be assigned to the classical glass transition in both polymers. By using different theoretical procedures, the curves for the loss modulus in the relaxation zones can be fitted and then interpreted. Figure 2.63 show an Arrhenius like plot which represent this behavior. The viscoleastic behavior of these systems is appropriately described using the classical fitting procedures used for similar systems and described previously [69].

According to the results shown for these polymers, the effect of the side chain structure on the viscoelastic and thermal behavior, play an important role. The effect of the carboxylic group by one hand and the length of the hydrophobic side chain on the other, are the driving forces responsible of the relaxational behavior in this family of poly(itaconate)s.

The effect of the side chain structure on the relaxational behavior of poly (itaconate)s was studied by Díaz Calleja and coworkers [238] in a family of poly(di-n-alkyl and diisoalkylitaconate)s. Specifically in poly(dimethyl itaconate) (PDMI), poly(diethylitaconate) (PDEI), poly(di-n-propyl itaconate) (PDPI), poly (di-n-butylitaconate) (PDIBI), poly(diisopropylitaconate) (PDIPI) and poly (diisobutylitaconate) (PDIBI). These systems show three dielectric relaxation processes, labelled as α, β and γ. Nevertheless, in some polymers a poor resolution of

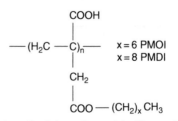

Scheme 2.15 Chemical structure of poly(monoitaconate)s. (From ref. [237])

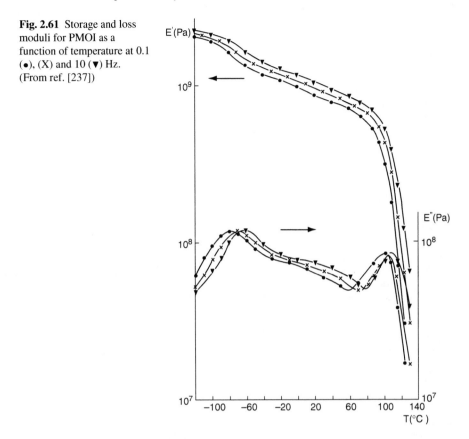

Fig. 2.61 Storage and loss moduli for PMOI as a function of temperature at 0.1 (●), (X) and 10 (▼) Hz. (From ref. [237])

the peaks is observed at low temperature. The mechanical behavior follows a similar behavior.

Figure 2.64. represents the storage modulus and the loss tangent at 1 Hz, for PDMI, PDEI, PDPI and PDBI as a function of temperature. Clearly the relaxation associated to the glass transition (Tg) is observed together with the effect of the length of the side chain on the relaxation process. The values of the temperature for $\tan \delta_{max}$ for the four polymers are 95°C, 71°C, 58°C and 30°C respectively. These values can be correlated with the corresponding values for the four first poly(meyhacrylate)s of the series (105, 74, 34, 18°C). Figure 2.65, show that the effect of the decrease of Tg with the side chain is more sensitive in the case of poly(methacrylate)s than in the case of poly(itaconate)s. It is indicative that in spite of the double esterification the puckering effect is more important in the case of poly(itaconate)s than in poly(methacrylate)s.

At low temperatures, cannot be detected viscoelastic activity defined in terms of loss tangent. Using the modulus E" Fig. 2.66 the results are better. Nevertheless the resolution remain weak although is higher than that reported by Cowie et al. [239–241].

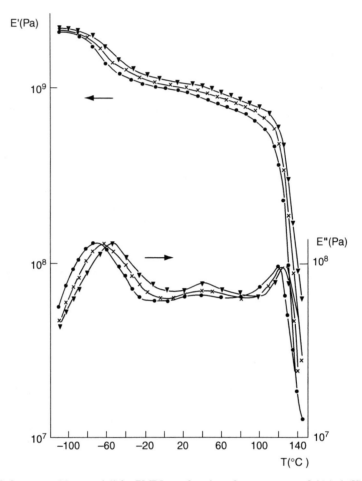

Fig. 2.62 Storage and loss moduli for PMDI as a function of temperature at 0.1(•), 1 (X) and 10 (▼). (From ref. [237])

For polymers containing branched side chains Fig. 2.67 show the variation of the modulus E', E" and the loss tangent for PDIPI and PDIBI in the temperature range under study. Two relaxations can be observed where the most prominent is the α relaxation associated to the glass transition as in the systems previously reported. Tg increases as the volume of the side chain increases. This result is in good agreement with that observed for the corresponding family of poly(methacrylate)s [242].

The results corresponding to dielectric relaxation measurements are more explicit than the mechanical ones. Figure 2.68 shows the data corresponding to ε' and tan δ for the four first members of the series of poly(itaconate)s.

The dielectric spectrum for polymers with bulky side chains are shown in Fig. 2.69 for PDIPI and PDIBI. In these cases beside the prominent α relaxations it is possible to observe conductive contributions at low frequencies and high temperatures. A relaxation map is summarized in Fig. 2.70.

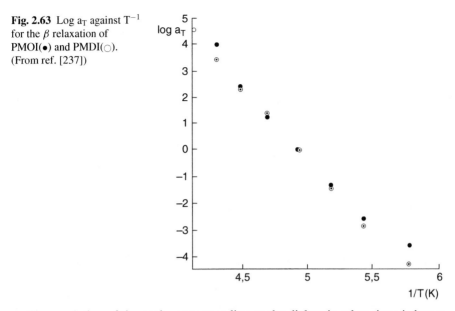

Fig. 2.63 Log a$_T$ against T^{-1} for the β relaxation of PMOI($\bullet$) and PMDI($\circ$). (From ref. [237])

The resolution of the peaks corresponding to the dielectric relaxations is better when the dielectric loss modulus formalism is used.

Figures 2.71 and 2.72 show conductivity contributions perhaps also overlapped by interfacial effects which are present for the peak at higher temperatures than that of the α relaxations. However there is not a clear correlation between the position of the peak and the length or shape of the side groups in these polymers.

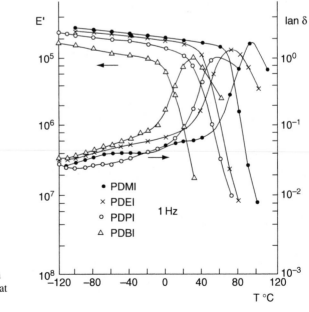

Fig. 2.64 Storage and loss tangent for ($\bullet$), PDMI, (X) PDEI($\circ$) and ($\triangle$) PDBI as a function of the temperature at 1 Hz. (From ref. [238])

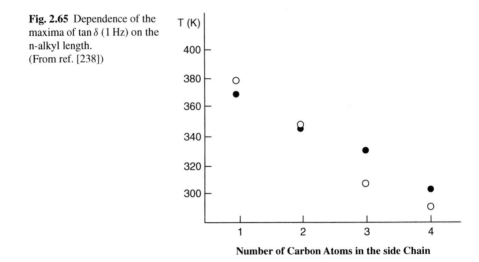

Fig. 2.65 Dependence of the maxima of tan δ (1 Hz) on the n-alkyl length. (From ref. [238])

Number of Carbon Atoms in the side Chain

That is an evidence of the spurious character of the contribution that can be attributed to free charges coming from impurities or the remainder of the solvent or initiator.

Figure 2.73 shows the plots for mechanical tan δ for PMMA, PDMI and PMMI. The first and third for comparison with the second. According to these results the β

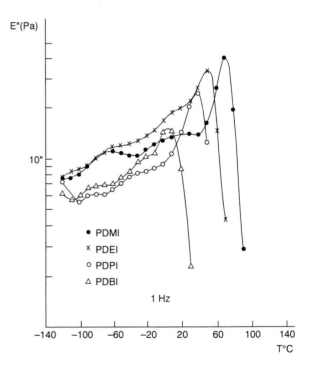

• PDMI
x PDEI
○ PDPI
△ PDBI

1 Hz

Fig. 2.66 Loss modulus for the same polymers under the same conditions as in Figure 2.4: (●) PDMI, (X) PDEI, (○) PDPI, (△) PDBI. (From ref. [238])

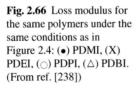

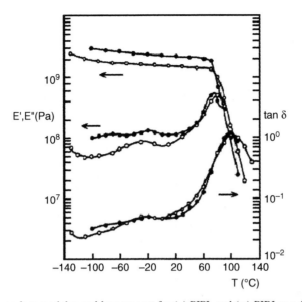

Fig. 2.67 Storage, loss modulus and loss tangent for (•) PIPI, and (○) PIBI as a function of the temperature at 1 Hz. (From ref. [238])

relaxation in PMMA persists in PDMI. This is the result of a more restricted motion than in the former or is probably overlapped by the α relaxation present in PDMI at lower temperatures than in PMMA.

The γ relaxation could be due to coordinated motions of the ester and acid groups. The dielectric results in terms of M" are shown in Fig. 2.74.

As a general comment about these systems, the molecular motions which would bw responsible for the sub. Tg relaxations in PDIPI and PDIBI can be due to the side chain where the spacer group –CH$_2$- is present, because there is no indication of the β relaxation that is associated with motions of the carboxyalkyl group directly joined to the main chain. This relaxation is always present, near room temperature. The values of the activation energies are in good agreement with the position of the peaks.

The presence of relaxations at very low temperatures such at that observed in Fig. 2.68 for PDPI and PDBI is in good agreement with that observed for poly(methacrylate)s [57].

Finally the δ relaxation in PDMI is found at lower temperatures than in PDPI. The long sequence of –CH$_2$- of the n-alkyl side groups give rise to a variety of degrees of freedom that can be the responsible for the relaxations observed. The potential barrier opposed in this case should be only intramolecular. Nevertheless the cooperativity due to the oxycarbonyl group cannot be disregarded in the motions which gives rise to the absorption. Motions in which more than one group play some role cannot be disregarded [238].

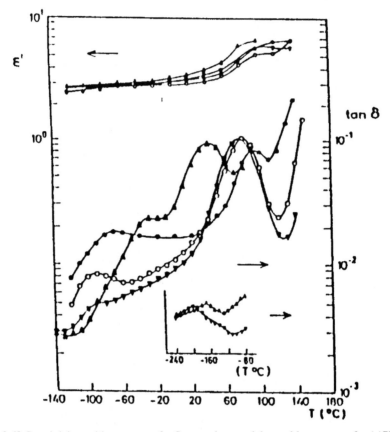

Fig. 2.68 Permittivity and loss tangent for Storage, loss modulus and loss tangent for (●)PDMI, (○) PDEI, (∇) PDPI, (△) PDBI. The variation of tan δ at lower temperatures for PDPI and PDBI is also shown at the bottom. Symbols are the same as those of the main figure. (From ref. [238])

2.5.2 Dielectric Relaxational Behavior

Poly(mono-cyclohexylitaconate) (PMCHI) (see Scheme 2.16) is a polymer that present several interesting characteristis. By one hand is a polymer containing a cyclohexyl group what, as it was discussed in previous sections, is a chemical structure that provide several relaxational responses due to the conformational versatility of the saturated cyclic side chain.

The dielectric behavior of PMCHI was studied by Díaz Calleja et al. [210] at variable frequency in the audio zone and second, by thermal stimulated depolarization. Because of the high conductivity of the samples, there is a hidden dielectric relaxation that can be detected by using the macroscopic dynamic polarizability α^* defined in terms of the dielectric complex permittivity ε^* by means of the equation:

$$\alpha^* = \frac{\varepsilon^* - 1}{\varepsilon^* + 2} \tag{2.45}$$

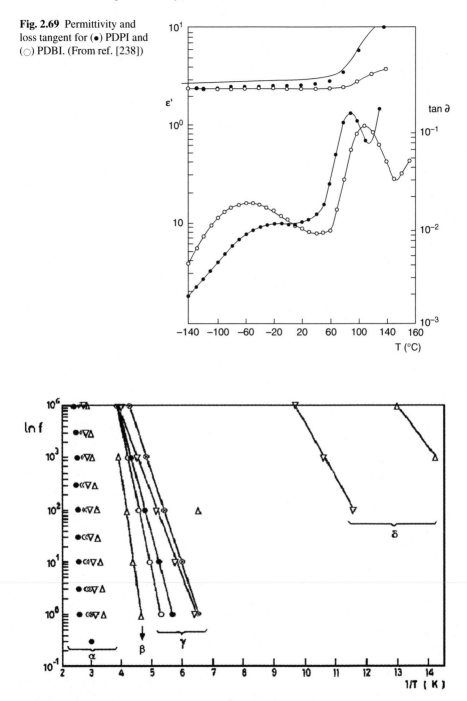

Fig. 2.69 Permittivity and loss tangent for (●) PDPI and (○) PDBI. (From ref. [238])

Fig. 2.70 Dielectric relaxation map, ln f vs T^{-1} showing α, β, γ and δ relaxational zones in the range of frequency and temperature studied: (●) PDMI, (○) PDEI, (∇) PDPI, ($\triangle$) PDBI, ($\otimes$) PDIBI. PDIPI is omitted for clarity of the figure. (From ref. [238])

Fig. 2.71 Dielectric loss
moduli for PDMI (●), PDEI
(X), PDPI (∇), and PDBI
(△)as function of
temperature. (From
ref. [238])

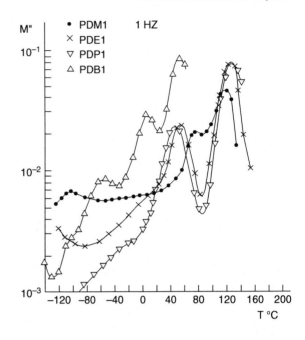

Fig. 2.72 Dielectric loss
moduli for PIPI (○) and PIBI
(●) as function of temperature
at 1 Hz. (From ref. [238])

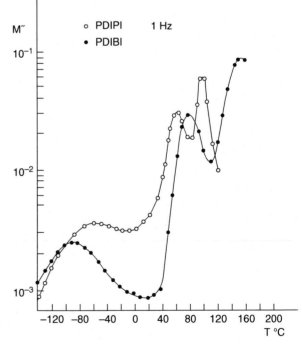

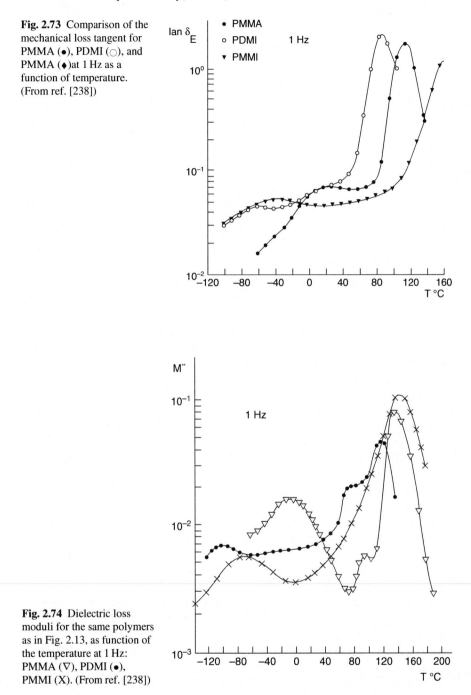

Fig. 2.73 Comparison of the mechanical loss tangent for PMMA (●), PDMI (○), and PMMA (♦)at 1 Hz as a function of temperature. (From ref. [238])

Fig. 2.74 Dielectric loss moduli for the same polymers as in Fig. 2.13, as function of the temperature at 1 Hz: PMMA (∇), PDMI (●), PMMI (X). (From ref. [238])

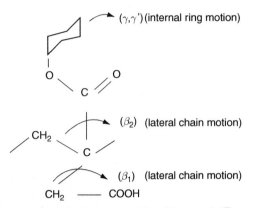

Scheme 2.16 Chemical structure of poly(monocyclohexyl itaconate). (From ref. [210])

where ε^* is the complex permittivity.

The transformation performed by this equation is a good way for the analysis of the dielectric relaxations in the zone of high temperatures and low frequencies of the spectrum.

Figure 2.75, shows the real and imaginary part of the dielectric permittivity at three characteristic frequencies two relaxation processes are clearly observed, one of them at low temperatures and the other al slightly lower room temperature. The second phenomenon is followed by an increasing of the loss due to conductivity, specially at low frequencies. Figure 2.76, shows the thermodepolarization spectra at two measurement conditions. The results in terms of $\tan \delta (= \varepsilon''/\varepsilon')$ are shown in Fig. 2.77, which reveals for the audio-frequency measurements a γ' relaxation process with an activation energy of 9 kcal mol^{-1} calculated by means of the Arrhenius equation from Ln υ_m vs T^{-1} where υ_m is the frequency of the maxima in $\tan \delta$. This relaxation (β) is overlapped by the next relaxation which is progressively broadening with increasing frequency. This behavior is similar to that found for other poly(monoitaconate)s [243–245].

At the same time, the activation energy calculated from the maxima in $\tan \delta$ increases from 28 to approximately 56 kcal mol^{-1}. This effect together with the asymmetric shape of the curves would indicate the presence of two sub relaxations, β_1 and β_2 in the increasing temperature sense.

The high conductivity observed at low frequencies would overlap the existence of another new relaxation (Fig. 2.78) what is very similar to that found for transfer complexes [243, 244] in which some of them have a pronounced semiconductor character. In this kind of compounds, the observation of dielectric relaxations over room temperature is inhibited by the high conductivity observed. To avoid this problem and to detect the conductivity effect, it is possible to use the complex polarizability α^* defined by equation [182]. The transformation defined by equation (2.45), has been applied with good results in the case of dielectric relaxation peaks in terms of α'' or $\tan \delta_\alpha$.

Fig. 2.75 ε' (*top*) and ε''
(*bottom*) for PMCHI ($\bullet$)
1 kHz; (X)10 kHz; ($\circ$)
100 kHz. (From ref. [210])

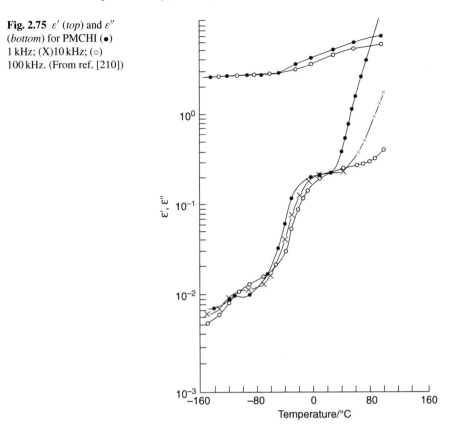

The variation of tan δ_α (referred to the polarizability α) as function of temperature is shown in Fig. 2.79. The activation energy value for this relaxation, calculated according to an Arrhenius equation for the five maxima, is 16 kcal mol^{-1}. This is a small value for a glass transition temperature, but not too much considering that in this case a small part of the macromolecule is activated from the dielectric point of view at higher temperatures than that of the β relaxation.

By comparing both series of results it is possible to conclude that the TSDC spectrum is notably more detailed, because it correspond to lower frequencies due to the greater splitting of the relaxation peaks. However both types of measurements can be considered as complementary. The technique used in charge-transfer complexes has been shown to be useful also in the case of polymers with high conductivity at high temperatures.

2.5.3 Dynamic Mechanical Relaxational Behavior

Figure 2.80 shows three main relaxations at different temperatures which are labeled as γ, β and α relaxations. The γ relaxation is attributed to the motions of

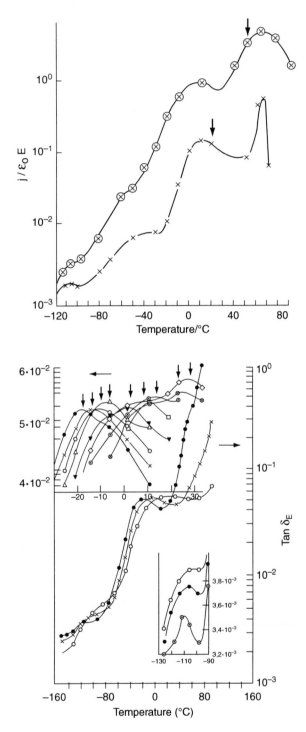

Fig. 2.76 Themally stimulated depolarization currents (TSDC). *Top*: (X); $T_{max} = 90°C$; $T_p = 50°C$; $E = 200\,V/mm$. *Bottom*: (X): $T_{max} = 70°C$; $T_p = 20°C$; $E = 150\,V/mm$. *Arrows* denote the polarization temperatures. (From ref. [210])

Fig. 2.77 Variation of $\tan \delta$ as function of temperature for PMCHI. *Top*: representation of the β relaxation: (•)0.2 kHz; (X) 0.5 kHz; (⊙) 1 kHz; (△) 2 kHz; (◊) 100 kHz. Middle: $\tan\delta_e$ for PMCHI at (•) 1, (X) 10, and (○) 100 kHz in the complete range of measurements. *Bottom*: Representation of the γ relaxation: (⊙) 0.5 kHz; (•) 1 kHz; (○) 2 kHz. (From ref. [210])

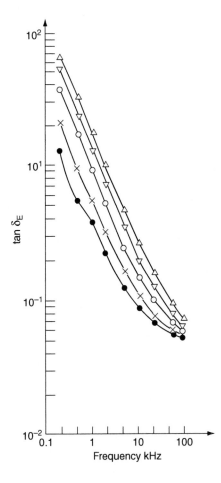

Fig. 2.78 Variation of tanδ_ε
as function of the frequency
for PMCHI: (●) 71°C; (X)
80°C; (○) 88°C; (◇) 95°C;
(△) 100°C. (From ref. [210])

the cyclohexyl group, specifically to the chair-to-chair flipping of the ring; and the
β relaxation to the rotation of the side carboxylic groups. In this case a mechanical model to analyze the α relaxation consider an isotropic continuum containing
spherical isotropic non-interacting particles. The peaks shown in Fig. 2.80, are
labeled as α, β and γ.

The γ relaxation at very low temperatures is a narrow mechanical absorption
with an activation energy of 13 kcal mol^{-1}.

At higher temperatures a new relaxation process appears. In this case the drop
in the storage modulus is higher, by nearly a decade, and from this point of view
this mechanical absorption has characteristic intermediate between an absorption
associated with a Tg and a secondary relaxation.

In the high temperature tail of the β relaxation and as a shoulder, there is evidence of a new relaxation which can be seen clearly at higher frequencies. As it was
previously reported, dielectric measurements carried out on the same polymer show
evidence of a new relaxation which cannot be observed directly, because of the high

Fig. 2.79 Variation of tan δ_α (referred to the polarizability α) as function of temperature: (•) 71°C; (X) 80°C; (○) 88°C; (△) 95°C; (∇)100°C. (From ref. [210])

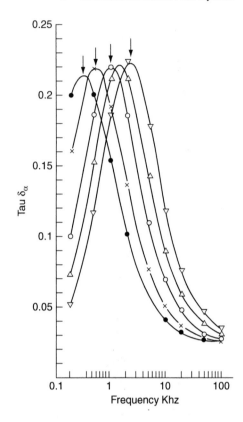

conductivity in the high-temperature zone of the spectrum. In order to make this phenomenon evident, a transformation from dielectric permittivity to polarizability has been demonstrated by Díaz Calleja et al. [209] to be a useful for this purposes. Mechanical relaxation measurements seem to be more clear in this respect [209] and a shoulder is observed in this case.

As a general comment about the dynamic mechanical relaxational behavior of this polymer, the results are consistent with dielectric data [210] and with the fact that no glass transition phenomenon is observed, at least in the range of temperature studied. This is striking in an amorphous polymer. It is likely that the residual part of the molecule mechanically active above the temperature of the β relaxation is only a small one, and this is the reason for the low loss observed in the α zone.

2.5.4 Dielectric Relaxational Behavior of Poly(diitaconate)s

Poly(dicyclohexyl itaconate)s

Polymers derived from itaconic acid show several relaxations below room temperature [245]. When these polymers contain cyclohexyl groups in the side chain like

Fig. 2.80 Storage (*upper curves*) and loss (*lower curves*) moduli of PMCHM at three frequencies: (•) 0.1, (X) 1, (○) 10 Hz. (From ref. [210])

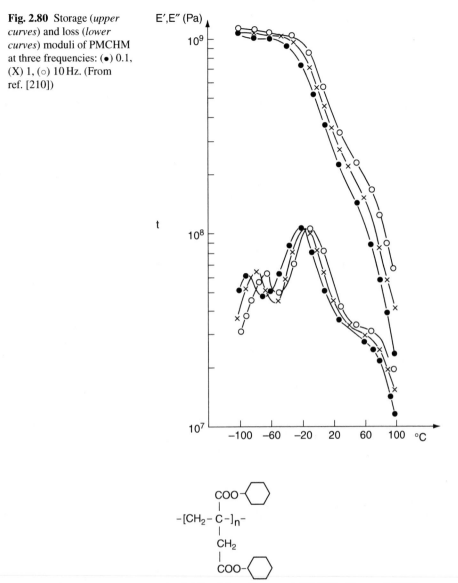

Scheme 2.17 Chemical structure of poly(dicyclohexylitaconate) (PDCHI). (From ref. [245])

poly(dicyclohexylitaconate) shown in Scheme 2.17, a prominent relaxation at about −75°C.

This is a typical relaxation for this kind of polymers [246, 247] and it has been attributed to the chair-to-chair conformational transition of the cyclohexyl ring like in the cases of poly(cyclohexyl methacrylate)s.

The dynamic loss modulus for PDCHI is shown in Fig. 2.81 where two relaxation processes can be observed at $-75°C$ and $-18°C$ at 1 Hz. which are labelled as γ and β.

From an Arrhenius type plot like that of Fig. 2.82 the activation energies for these processes are obtained an the values are 14.6 and 57.0 kcal mol^{-1}. these results give some idea about the differences in sizes of groups involved in the relaxation.

To obtain information about both relaxations, the Eyring equation, from the theory of absolute reaction rates, for the dependence of the frequency of an absorption peak on temperature has been used:

$$f = \frac{kT}{2\pi h} \exp\left(-\frac{\Delta G}{RT}\right) \qquad (2.46)$$

where k and h are the Boltzmann and Plank constants R is the gas constant and ΔG is the Gibbs free energy of the barrier to the relaxation process. Therefore from the classical relation:

$$\Delta G = \Delta H - T\Delta S \qquad (2.47)$$

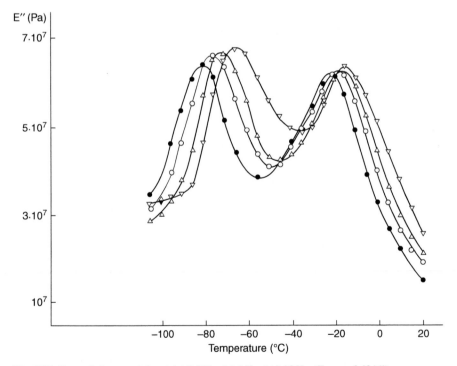

Fig. 2.81 Dynamic loss modulus at ($\bullet$) 0.3 Hz; ($\circ$) 1 Hz; ($\triangle$) 10 Hz. (From ref. [246])

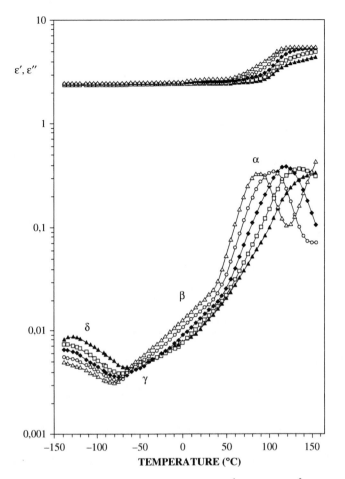

Fig. 2.82 Dielectric permittivity and loss for PDCBI (▲) 5×10^4 Hz, (□) 5×10^3 Hz, (♦) 5×10^2 Hz, (○) 5×10^1 Hz, and (△) 5×10^0 Hz. (From ref. [60])

It is possible to obtain:

$$\ln f = \ln \frac{k}{2\pi h} + \frac{\Delta S}{R} - \frac{\Delta H}{RT} \tag{2.48}$$

The experimental results show that the activation entalphy of the β relaxation is for times that of the γ relaxation and the activation entropy is ten times greater. Starkweather has described the secondary relaxation a simple or complex according to the value of the activation entropy. In general there is low cooperativity in the former relative to the latter. Motions of small groups at cryogenic temperatures are examples of the first type of relaxation while some β relaxations involving the participation of the main chain are representative of the second.

poly(diitaconate)s Containing Different Cyclic Side Chains

The dielectric relaxational behavior of several poly(diitaconate)s containing cyclic
rings in the side chain (see Scheme 2.18) show different behaviors at low temperatures depending on the chemical structure of the polymers.

Figures 2.82, 2.83, and 2.84 illustrate the dielectric permittivity and loss for PD-
CBI, PDCHpI and PDCOI at different frequencies. The α relaxation associated to
the glass transition is clearly observed in these Figures. The β relaxation is also
observed as a shoulder in the low temperature side of the α relaxation. Moreover,
γ and δ relaxations are also present depending on the structure of the polymer.
Particularly, for PDCHpI only a weak subglass activity is observed in the low range
of temperatures.

According to the results reported in these works, polymer with even-membered
rings display more prominent relaxational activity than those with odd-membered
rings.

Assuming that the subglass dielectric activity observed in these poly(diitaconate)s
have a similar origin than that observed in mechanical measurements by Heijboer
[28] the results for these systems are in good agreement with those previously
reported.

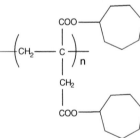

PDCpHpI

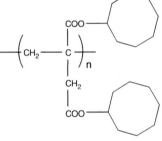

PDCOI

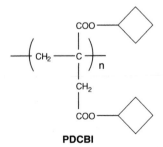

PDCBI

Scheme 2.18 Chemical structures of poly(dicyclo butyl, heptyl and octyl itaconate)s. (From
ref. [60])

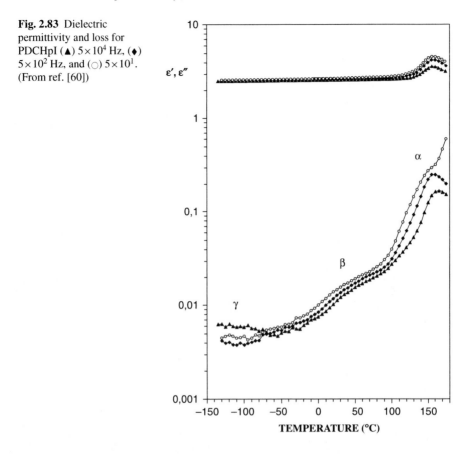

Fig. 2.83 Dielectric permittivity and loss for PDCHpI (▲) 5×10^4 Hz, (♦) 5×10^2 Hz, and (○) 5×10^1. (From ref. [60])

In the case of PDCHI, measurements at lower temperatures allow to visualize more clearly the δ relaxation as shown in Fig. 2.85. In this case a new relaxational process appears, which is labeled as γ relaxation.

γ and δ relaxations are attributed to the motions in the four-membered ring, without excluding the possibility of some type of cooperation of the ester group. Therefore it is possible to speculate that the butterfly motions inside the four-membered ring are involved in these two processes.

The γ relaxation for PDCOI can be analyzed using the classical procedures. According to previous research [28], the eight-membered ring has a prominent γ mechanical loss that is higher than the six, ten-, and twelve-membered ring analogues. A large number of conformations is possible for cyclooctane ring. Two of the following groups of these conformations are particularly important: the boat-to-chair family and the corona family, to which also belong the chair-to-chair and twisted chair-to-chair conformations. The mechanical loss peak is attributed [28] to some type of boat-to-chair inversion or boat-to-chair to twist chair-to-chair interconversion. Moreover, the activation energy of the mechanical γ relaxation of poly(cyclooctyl methacrylate) is 44.3 kJ mol^{-1} which is very

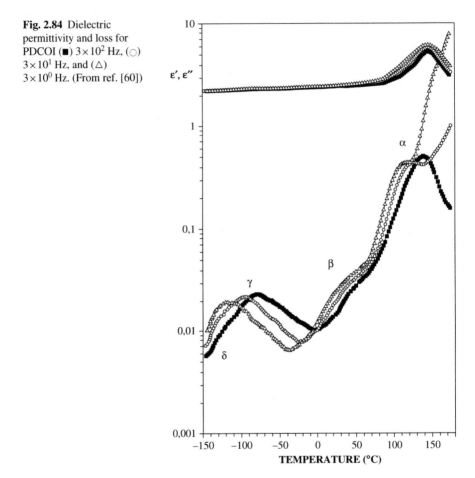

close to the result obtained for the dielectric γ relaxation of PDCOI. For this reason it is possible to conclude that the conformational transition inside the cyclooctane ring are responsible for the γ -dielectric relaxation of PDCOcI.

A β relaxation is observed as a shoulder of the α relaxation in all the polymers under study in this section. This type of β relaxation is commonly attributed to rotational motions of the lateral chain as a whole as in poly (alkyl methacrylates) [34–40, 60, 61].

2.6 Viscoelastic Properties of Poly(thiocarbonate)s

As a general example of another systems that show interesting responses from relaxational point of view, condensation polymers like poly(carbonate)s and poly (thiocarbonate)s are a new group of polymers to be analyzed. Poly(thiocarbonate)s

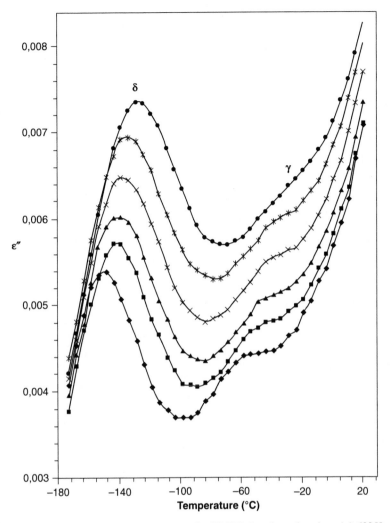

Fig. 2.85 Loss permittivity at low temperature for PDCBI: δ and γ relaxations (♦) 620 Hz, (■) 1820 Hz, (▲)2610 Hz, (X) 3750 Hz, and (●) 12.000 Hz. (From ref. [60])

(PTC) is a family of polymers whose thermal properties, Tg, unperturbed dimensions and partial specific volumes has been reported [248–252]. The dielectric properties of these polymers were recently studied [253–256]. Relaxational studies on poly(thiocarbonate)s are scarce, but on the contrary there is much information about relaxation processes of the analogues poly(carbonate)s (PC). Therefore the study of PTC is an interesting approach to get confidence about the motions responsible of the relaxational behavior of these polymers.

One of the first PTC studied from relaxational point of view was poly[2,2-propane-bis-(4-phenylthiocarbonate)] [257]. The chemical structure of PTC is similar to that of PC, (see Scheme 2.19) therefore the starting hypothesis is to assign the origin of the molecular motions to these polymers in a similar way to that of PC's.

The dielectric relaxation properties of poly[2,2-propane-bis-(4-phenylthiocarbonate)] have been studied by Diaz Calleja et al. [257] and the crystallinity of the samples is considered as a trouble which can be eliminated by quenching. The degree of crystallinity of the polymers is determined by differential scanning calorimetry in order to investigate the effect of crystallinity on the dielectric behavior. The degradation of the polymer began before the glass transition temperature Tg. The dielectric spectrum is complex showing several relaxation phenomena. With increasing temperature a γ relaxation is observed at $-100°$C (5 kHz). The activation energy obtained from an Arrhenius plot of lnf vs T^{-1} is 6 kcal mol^{-1}. At 160°C the α relaxation associated to the glass transition temperature (Tg) is clearly detected. The comparison of the dielectric behavior of these polymers with those of PCs is a good way to understand the origin of the molecular motions of these polymes.

Important contributions to the effect of the substituents in the carbon atom between both phenyl rings i.e. the interphenylic carbon atom was reported by Sundararajan [258–260], Bicerano and Clark [261,262], as well as Hutnik and Suter [263, 264]. Diaz Calleja et al. [265] have reported the dielectric relaxational behavior of a family of poly(thiocarbonate)s with the basic chemical structure shown in Scheme 2.19.

The variation of $tan\delta$ with temperature at 1 kHz for the six poly(thiocarbonate)s is represented in Fig. 2.86. In all cases a prominent relaxation associated to the glass transition temperature labelled as α -relaxation is observed in Figure PT-1. A secondary relaxation which covers a range of about hundred degrees and which by comparison with the results reported for PCs is labeled as γ relaxation. Between 80°C and 100°C a slightly dielectric activity is observed (β zone) and at $-120°$C another relaxation labelled as δ relaxation for polymers 4,5 and 6.

As a general conclusion, poly(thiocarbonate) behaves in a similar way to poly-(carbonate)s in particular a γ relaxation appears in the same range of temperatures in both type of polymers.

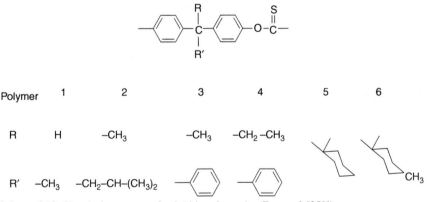

Scheme 2.19 Chemical structures of poly(thiocarbonate)s. (From ref. [258])

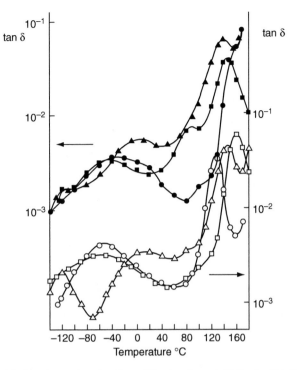

Fig. 2.86 Dielectric loss tangent as function of temperature at 1 kHz for the six polymers of Scheme 2.19. (From ref [258]) ○, 1; ●, 2; □, 3; ■ 4; △, 5; ▲,6. (From ref. [258])

As Diaz Calleja pointed out [258], physical aging also affects the dielectric properties between room and higher and lower temperatures. Conductivity contributions to the loss permittivity are observed in nearly all the polymers under study.

References

1. N.G. McCrum, C.P. Buckley, C.B. Bucknall, "Principles of Polymer Engineering", Oxford University Press (1995).
2. Van Vliet, Krystyn J. "Machanical Behaviour of Materials" 3.032 (2006).
3. M.A. Meyers, K.K. Chawla "Mechanical Behavior of Materials" 3.032 (2006).
4. The Open University, Learning Space. http://openlearn.open.ac.uk/mod/resource/view.php?id= 196619 (1008).
5. E. Riande, R. Díaz-Calleja, M.G. Prolongo, R.M. Masegosa, C. Salom, "Polymer Viscoelasticity" Marcel Dekker, New York –Basel (2000).
6. J.D. Ferry, "Viscoelastic Properties of Polymers" 3rd ed. New York: Wiley-Interscience (1980).
7. A. Horta, "Macromoléculas" Universidad Nacional de Educación a Distancia", 2, Madrid (1982).
8. Wikipedia Encyclopedia "Viscoelasticity" (2008 for clinical applications).
9. E. Riande, R. Díaz-Calleja, Prolongo, R.M. Masehosa, C. Salom, "Polymer Viscoelasticity", Marcel Dekker, Inc., New York (2000).
10. D.J. Plazek, J. Phys. Chem., 69, 3480 (1956).

11. E. Riande, H. Markovitz, D.J. Plazek, N. Raghupati, J. Polym. Sci., C50, 405 (1975).
12. J.J. Aklonis, W.J. MacKnight, "Introduction to Polymer Viscoelasticity", 2nd ed., New York: Wiley (1980).
13. T.G. Fox, J.P. Flory, J. Appl. Phys., 21, 581 (1950).
14. P.J. Flory, "Principles of Polymer Chemistry", New York: cornell University Press (1953).
15. M. Yazdani-Pedram, E. Soto, L.H. Tagle, F.R. Díaz, L. Gargallo, D. Radić, Themochimica Acta 105, 149 (1986).
16. L. Gargallo, N. Hamidi, D. Radić, Thermochim. Acta, 114, 319 (1987).
17. L. Gargallo, N. Hamidi, L.H. Tagle, D. Radić, Thermochim. Acta, 124, 263 (1988).
18. L. Gargallo, M.I. Muñoz, D. Radić, Thermochim. Acta 146, 137 (1989).
19. A. Ribes-Greus, R. Díaz-Calleja, L. Gargallo, D. Radić, Polym. Intern., 25, 51 (1991).
20. R. Diaz-Calleja, A. García Bernabé, E. Sanchez-Martínez, A. Hormazábal, L. Gargallo, D. Radić, Macromolecules, 34, 6312 (2001).
21. G. Domíguez-Espinoza, R. Díaz-Calleja, E. Riande, L. Gargallo, D. Radić, Macromolecules 39, 3071 (2006).
22. G. Domínguez-Espinosa, R. Díaz-Calleja, E. Riande, L. Gargallo, D. Radić, J. Chem. Phys., 123 114904 (2005).
23. R. Silbey, R. Alberty M. Bawendi (2001), "Physical Chemistry", 4th ed. John Wiley & Sons, Inc (2006).
24. Allen, Thomas "The Structure of Materials," 51 (1999).
25. D. Roylance, Report, Department of Material Science and Engineering, Massachusetts Institute of Technology, Cambridge, MA 02139, October 24 (2001).
26. K. Onaram, W.H. Findley, "Creep and Relaxation of Nonlinear Viscoelastic Materials", Dover Publications, New York (1989).
27. G. Dominguez-Espinosa, R. Díaz-Calleja, E. Riande, L. Gargallo, D. Radić, Macromolecules 39, 3071 (2006).
28. J. Heijboer, Ph.D. Thesis, Leiden (1972).
29. R. Diaz-Calleja, L. Gargallo, D. Radić, Polymer, 34, 4247 (1993).
30. A. Ribes Greus, R. Díaz-Calleja, L. Gargallo, D. Radić, Polymer, 30, 1685 (1989).
31. R. Díaz-Calleja, M.J. Sanchis, E. Saiz, F. Martínez-Piña, R. Miranda, L. Gargallo, D. Radić, E. Riande, J. Polym. Sci., Part B, Polymer Physics, 38, 2179 (2000).
32. R. Díaz-Calleja, A. García Bernabé, E. Sanchez-Martínez, A. Hormazábal, L. Gargallo, D. Radić, Polymer, 41, 4811 (2000).
33. R. Diaz-Calleja, A. García Bernabé, E. Sanchez-Martínez, A. Hormazábal, L. Gargallo, F. González-Nilo, D. Radić, Macromolecules, 34, 6312 (2001).
34. R. Diaz-Calleja, A. García Bernabé, C. Pagueguy, L. Gargallo, D. Radić, Polym. Inter., 51, 808 (2002).
35. R. Diaz-Calleja, A. García Bernabé C. Pagueguy, L. Gargallo, D. Radić, Polymer 42, 8907 (2001).
36. R. Diaz-Calleja, A. García Bernabé, E. Sanchez-Martínez, A. Hormazábal, L. Gargallo, D. Radić, Polym. Inter., 51, 808 (2002).
37. R. Diaz-Calleja, A. García Bernabé, E. Sanchez-Martínez, A. Hormazábal, L. Gargallo, D. Radić, Polym. Inter. 51, 1448 (2002).
38. M.J. Sanchis, G. Dominguez-Espinoza, R. Diaz-Calleja, A.L. Alegría, L. Gargallo, D. Radić, J. Polym. Sci., B Polym. Phys., 44, 3135 (2006).
39. R. Diaz-Calleja, A. García Bernabé, L. Gargallo, D. Radić, Intern. J. Polym. Mater., 55, 1 (2006).
40. O. Pelissou, R. Díaz-Calleja, L. Gargallo, D. Radić, Polymer 35, 3449 (1994).
41. R. Díaz-Calleja, L. Gargallo, D. Radić, Polym. Bull., 32, 675 (1994).
42. R. Díaz-Calleja, I. Devine, L. Gargallo, D. Radić, Polymer 35, 151 (1994).
43. R. Díaz-Calleja, C. Jaime's, M.J. Sanchis, F. Martínez-Piña, L. Gargallo, D. Radić, Polym. Eng. Sci., 37, 882 (1997).

44. R. Díaz-Calleja, C. Jaime's, M.J. Sanchis, J. San Roman, L. Gargallo, D. Radić, Macromol. Chem. Phys. 199, 575 (1988).
45. M.J. Sanchis, R. Díaz-Calleja, A. Hormazábal, L. Gargallo, D. Radić Macromolecules 32, 3457 (1999).
46. R. Díaz-Calleja, M.J. Sanchis, E. Saiz, F.Martínez-Piña, R. Miranda, L. Gargallo, D. Radić, E. Riande, J. Polym. Sci. Polym. Phys. Ed., 38, 2179 (2000).
47. R. Díaz-Calleja, A. García Bernabé, E. Sánchez Martínez, A. Hormazábal, L. Gargallo, D. Radić, Polymer, 41, 4811 (2000).
48. R. Díaz-Calleja, A. García Bernabé, E. Sánchez Martínez, A. Hormazábal, L. Gargallo, F. González Nilo, D. Radić, Macromolecules 34, 6312 (2001).
49. R. Díaz-Calleja, A. Garcia-Bernabé, C. Pagueguy, L. Gargallo, D. Radić, Polym. Commun. 42, 8907 (2001).
50. R. Díaz-Calleja, A. García-Bernabé, E. Sánchez-Martínez, A. Hormazábal, L. Gargallo, D. Radić, Polym. Intern. 51, 1448 (2002).
51. G. Domínguez, M.J. Sanchis, R. Díaz-Calleja, C. Pagueguy, L. Gargallo, D. Radić, Polymer 46, 8028 (2005).
52. G.P. Mikhailov, T.I. Borisova, Vysokomol. Soedin., 6, 1778 (1964).
53. G.P. Mikhailov, T.I. Borisova, Vysokomol. Soedin., 6, 1784 (1964).
54. K. Shimizu, O. Yano, Y. Wada, J. Polym. Sci., 13, 1959 (1964).
55. M. Kakisaki, K. Aoyama, T. Hideshima, Rep. Prog. Polym. Sci. Jpn., 17, 379 (1974).
56. D.J. Mead, R.M. Fuoss, J. Am. Chem. Soc. 64, 2389 (1942).
57. J. Heijboer, M. Pineri in "Non-Metallic Materials and Composites at Low Temperatures 2", Plenum, New York, p.89 (1982).
58. J.L. Gómez Ribelles, R.Díaz Calleja, J. Polym. Sci., 23, 1565 (1985).
59. R. Díaz-Calleja, A. Garcia-Bernabé, A. Hormazábal, L. Gargallo, D. Radić, Polym. Int., 51, 808 (2002).
60. R. Díaz-Calleja, A. García Bernabé, E. Sánchez Martínez, A. Hormazábal, L. Gargallo and D. Radić, J. Polym. Sci., Polym. Phys. Ed. 41, 1059 (2003).
61. G. Domínguez, M.J. Sanchis, R. Díaz-Calleja, C. Pagueguy, L. Gargallo, D. Radić, Polymer 46, 11351 (2005).
62. G Dominguez-Espinosa, R. Díaz-Calleja, E. Riande, L. Gargallo, D. Radić, J. Chem. Phys., 123, 114904 (2005).
63. R. Díaz-Calleja, A. García-Bernabé, L. Gargallo, D. Radić, Int. J. Polym. Mater., 55, 1155 (2006).
64. M.J. Sanchis, R. Díaz-Calleja, A. Garcia Bernabé, L. Alegría, L. Gargallo, D. Radić, J. Polym. Sci. Polym. Phys. 46, 109 (2008) and references therein.
65. K. Pathmanathan, G.P. Johari, J. Chem. Phys., 95, 5990 (1992).
66. N.G. McCrum, B. Read, G.Williams, "Anelastic and Dielectric Effects in Polymeric Solids", Dover New York, 1991, 1967, p. 24.
67. R. Díaz-Calleja, E. Riande, J. San Román, V, Compañ, Macromolecules, 27, 2092 (1994).
68. E. Saiz, E. Riande, J. Chem. Phys., 103, 3832 (1995).
69. R.M. Fuoss, J.G. Kirkwood, J. Am. Chem. Soc., 63, 385 (1941).
70. S. Havriliak, S. Negami, J. Polym. Sci., Polym., Symp., 14, 99 (1966).
71. T.S. Sorensen, V. Compañ, R. Díaz Calleja, J. Chem. Soc., Faraday Trans, 92, 1947 (1996).
72. T. Ojeda, D. Radić, L. Gargallo, Makromol. Chem., 181, 2237 (1980).
73. L. Gargallo, M.I. Muñoz, D. Radić, Polymer, 27, 1416 (1986).
74. J. Niezette, N. Hadjichristidis, V. Desreux, Makromol. Chem., 177, 2069 (1974).
75. M. Becerra, L. Gargallo, D. Radić, Makromol. Chem., 179, 2241 (1978).
76. F. Martínez-Piña, L. Gargallo, A. Leiva, D. Radić, Int. J. Polym. Anal. Charact. 4, 31 (1997).
77. D. Radić, F. Martínez-Piña, L. Gargallo, N. Hamidi, Polym. Eng. Sci., 36, 188 (1996).
78. L. Gargallo, I. Méndez, D. Radić, Makromol. Chem., 184, 1953 (1984).
79. L. Gargallo, J. Niezette, V. desreux, Bull. Soc. R. Sci. Liège, 46, 82 (1977).
80. N. Hadjichristidis, M. Devaleriola, V. Desreux, Eur. Polym. J., 8, 1193 (1972).

81. E. Saiz, A. Horta, L. Gargallo, I. Hernández-Fuentes, D. Radić, Macromolecules, 21, 1376, (1988).
82. R. Díaz-Calleja, E. Saiz, E. Riande, L. Gargallo, D. Radić, Macromolecules 26, 3795 (1993).
83. R. Díaz-Calleja, L. Gargallo, D. Radić, Polymer, 33, 1406 (1992).
84. Díaz-Calleja, L. Gargallo and D. Radić, Polymer, 33, 1406 (1992).
85. R. Díaz-Calleja, E. Sánchez-Martínez, A. Hormazábal, F. Martínez-Piña, L. Gargallo, D. Radić, Polym. Int., 51, 808 (2002).
86. D. Boese, S. Kremer, Macromolecules, 23, 829 (1990).
87. S. Havriliak, S. Negami, Polymer, 8, 161 (1967).
88. H. Vogel, Z. Phys., 22, 645 (1921).
89. G.S. Fulcher, J. Am. Ceram. Soc., 8, 339 (1925).
90. G. Tamman, W. Hesse, Z. Anorg. Allg. Chem., 156, 245 (1926).
91. R.M. Fuoss, J.G. Kirkwood, J. Am., Chem., Soc., 63, 385 (1941).
92. J.R. McDonald, Phys. Rev., 92, 4 (1953).
93. R. Coelho, Rev. Phys. Appl., 18, 13 (1983).
94. R. Díaz-Calleja, A. García Bernabé, E. Sánchez Martínez, A. Hormazábal, L. Gargallo, D. Radić, J. Polym. Sci., Polym. Phys. Ed. 41, 1059 (2003).
95. R. Díaz-Calleja, F. Martínez-Piña, L.Gargallo, D. Radić, Polym. Eng. Sci., 39, 882 (1997).
96. A. Ribes-Greus, J.L. Gómez Ribelles, R. Díaz-Calleja, Polymer, 26, 1849 (1985).
97. J. Heijboer, Z. Kolloid, 148, 36 (1956).
98. J. Heijboer, J.M.A. Baas, B. Van Graaf, M.A. Hoefnagel, Polymer, 28, 509 (1987).
99. J. Heijboer, J.M.A. Baas B. Van Graaf, M.A. Hoefnagel, Polymer, 33, 1359 (1992).
100. G. Dominguez-Espinoza, M.J. Sanchis, R. Díaz-Calleja, C. Pagueguy, L. Gargallo, D. Radić, Polymer, 46, 8028 (2005).
101. G. Dominguez-Espinoza, M.J. Sanchis, R. Díaz-Calleja, C. Pagueguy, L. Gargallo, D. Radić, Polymer, 46, 11351 (2005).
102. R. Díaz-Calleja, M.J. Sanchis C. Alvarez, E. Riande, J. Appl. Phys., 80, 1047 (1997).
103. R.Díaz-Calleja, M.J. Sanchis, C. Alvarez, E. Riande, J. Appl. Phys., 81, 3685 (1997).
104. N. García, V. Compañ, R. Díaz-Calleja, J. Guzman, E.Riande, Polymer, 41, 6603 (2000).
105. M.J. Sanchis, G. Domínguez-Espinoza, R. Díaz-Calleja, L. Alegría, L.Gargallo, D. Radić, J. Polym. Sci., Part B. Polym. Phys., 44, 3135 (2006).
106. T. Philip, R.L. Cook, T.B. Malloy, N.L. Allinger, S. Chang, Y. Yuh, J. Am. Chem. Soc., 103, 2151 (1981).
107. N.L. Allinger, S. Chang, Y. Yuh, Quant. Chem., Program, Exch., 13, 395 (1981).
108. N.L. Allinger, X. Zhou, J. Bergsma, J. Mol. Struct (Theochem) 118, 69 (1994).
109. U. Burkert, N.L. Allinger, ACS Monographs Series, Number 177: Molecular Mechanics ACS: Washington, DEC, (1982).
110. Serena Software, PC-MODEL. IN: Serena Software Bloomington.
111. N.L. Allinger, J.T. Sprague, J. Am. Chem. Soc., 95, 3893 (1973).
112. J.M.G. Cowie, I.J. McEwen, Macromolecules, 14, 1374 (1981).
113. J.M.G. Cowie, I.J. McEwen, Macromolecules, 14, 1378 (1981).
114. J.M.A. Baas, B. van Graaf, B. Heijboer, Polymer, 32, 2141 (1991).
115. K. Shimizu, O. Yano, Y. Wada, J. Polym. Sci., 13, 1959 (1975).
116. C. Esteve-Marcos, M.J. Sanchis, R. Díaz-Calleja, Polymer, 38, 3805 (1997).
117. M.J. Sanchis, R. Diaz-Calleja, O. Pelissou, L. Gargallo, D. Radić, Polymer, 45, 1845 (1997).
118. A. De la Rosa, L. Heux, J.Y. Cavaille', K. Mazeau, Polymer, 43, 5665 (2002).
119. M.S. Sulhata, U. Natajaran, Macromol. Theory Simul., 12, 61 (2003).
120. J.M.G. Cowie, R. Ferguson, Polymer, 28, 503 (1987).
121. S.C. Kuebler, D.J. Schaefer, C. Boeffel, U. Pawelzik, H.W. Spiess, Macromolecules, 30, 6597 (1997).
122. J.D. Ferry, "Viscoelastic Properties of Polymers", Wiley Interscience, New York, (1980).
123. M.P. Patel, M. Branden, Biomaterials 12, 645 (1991).
124. M.P. Patel, M. Branden, Biomaterials 12, 649 (1991).
125. M.P. Patel, M. Branden, Biomaterials12, 653 (1991).

126. M.P. Patel, M. Branden, S.J. Downes, J. Mater. Sci. Mater. Med., 5, 338, (1994).
127. L. Di Silvio, M.V. Kayser, S.M. Mirza, V. Downes, Clin. Mater. 16, 91 (1998).
128. M.P. Patel, G.J. Pearson, M. Branden, M.A. Mirza, Biomaterials, 19, 1911 (1998).
129. P.D. Riggs, M. Braden, M. Patel, Biomaterial, 21, 245 (2000).
130. S. Downes, R.S. Archer, M. Kayser, M.P. Patel, M. Branden, J. Mater. Sci. Mater. Med., 5, 88 (1994).
131. N. Reissis, S. Downes, M.V. Kayser, D. Lee, G. Bentley, J. Mater Sci. Mater. Med., 5, 402 (1994).
132. N. Reissis, S. Downes, M.V. Kayser, D. Lee, G. Bentley, J. Mater Sci. Mater. Med., 5, 793 (1994).
133. C.D. McFarland, S. Mayer, C. Scotchford, B.A. Dalton, J.G. Steele, S.J. Downes, J. Biomed. Matter. Res, 44, 1 (1999).
134. R.M. Wyre, S. Downes, Biomaterials, 21, 335 (2000).
135. J.C. Ronda, A. Serra, V. Cádiz, Macromol. Chem. Phys., 199, 343 (1998).
136. P.D. Riggs, M. Braden, T.A. Tibrook, H. Swai, R.L. Clarke, M. Patel, Biomaterials, 20, 435 (1999) and references therein.
137. M.P. Patel, H. Swai, K.W.M., Davy, M. Braden, J. Mater. Sci. Mater. Med., 10, 147 (1999).
138. R.M. Sawtel, S. Downes, R. Clake, M. Patel, M. Braden J. Mater. Sci. Mater. Med., 8, 667 (1997).
139. G.J. Pearson, D.C.A. Picton, M. Branden, C. Longman., Int. Endood. J., 19, 121 (1986).
140. S.J. Velickovíć, M.T. Kalagasidis Krusíć, R.V. Pjanovíć, N.M. Boskovíć-Vragolovíć, P.C. Griffiths, I.G. Popovíć, Polymer, 46, 7982 (2005).
141. A.K. Jonscher, Phil. Mag., B, 38, 587 (1978).
142. M. Sun, S. Pejanovíć, J. Mijovíć, Macromolecules, 38, 9854 (2005).
143. J. Mijovíć, S. Ristíć, J. Kenny, Macromolecules, 40, 5212 (2007).
144. S. Mashimo, S. Kuebara, S. Yagihara, K. Higasi, J. Phys. Chem., 91, 6337 (1987).
145. L.A. Dissado, R. Hill, J. Chem. Soc., Faraday, Trans, 80, 291 (1984).
146. I.M. Hodge, K. Ngai, C.T. Moinihan, J. NonCryst. Solids, 351, 104 (2005).
147. K.W. Wagner, Arch. Electrotechnol., 2, 371 (1914).
148. K.W. Wagner, Arch. Electrotechnol., 3, 67 (1914).
149. R.W. Sillars, Proc. Inst Electr. Eng. Lond., 80, 378 (1937).
150. R. Bergman, J. Swenson, Nature 403, 283 (2000).
151. R. El Moznine, G. Smith, E. Polygalov, P.M. Suherman, J. Broadhead, J. Phys. D. Appl. Phys., 36, 330 (2003).
152. J. Pérez, J.Y. Cavaillé, T.J. Tatibouet, J. Chim., Phys., 87, 1993 (1990).
153. R. Díaz Calleja, E. Riande, J. Guzman, J. Makromol. Chem. 25, 1326 (1990).
154. R. Díaz Calleja, E. Riande, J. San Roman, Macromolecules 24, 1854 (1991).
155. B.B. Bauer, P. Avakian, H.W. Starkweather Jr., B.S. Hsiao, Macromolecules 23, 5119 (1990).
156. Y. Kihira, K. Sugiyama, Makromol. Chem., 187, 2445 (1986).
157. J.S. Velickovíć, J.M. Filipovíć, M.B. Plasvíć, D.M. Petrovíć-Djacov, Z.S. Petrovíć, J.S. Budinski, Polym. Bull. 27, 331 (1991).
158. R. Díaz-Calleja, E. Riande, J. San Roman, Polymer 32, 2995 (1991).
159. C. Huet, Ann. Ponts. Chaus V, 5 (1965).
160. J. Pérez, J.Y. Cavaillé, S. Etienne, C. Jourdan, Rev. Phys. Appl., 23, 125 (1988).
161. J.Y. Cavaillé, J. Pérez, G.P. Johari, Phys. Rev. J. Non-Cryst Solids, 131, 935 (1991).
162. R. Díaz-Calleja, E. Riande, J. Guzman, J. Phys. Chem., 96, 932 (1992).
163. R. Díaz-Calleja, E. Riande, J. Guzman, J. Phys. Chem., 95, 7104, (1991).
164. G.D. Smith, R.H. Boyd, Macromolecules, 25, 1326 (1992).
165. H.W. Starkweather Jr., Macromolecules, 23, 328 (1990).
166. S. Koizumi, K. Tadano, Y. Tanaka, T. Shimidzu, T. Kutsumizu, S. Yano, Macromolecules, 25, 6563 (1992).
167. S.-Y. Park, S.N. Chavalum, J. Blackwell, Macromolecules, 30, 6814 (1997).
168. G.U. Pittman, M. Ueda, K. Iri, Y. Imai, Macromolecules, 13, 1031 (1980).

169. Y. Okawa, T. Maekawa, Y. Ishida, M. Matsuo, Polym. Prepr. Jpn, 40, 3098 (1991).
170. K. Tadano, Y. Tanaka, T. Shimisu, S. Yano, Macromolecules, 32, 1651 (1999).
171. N.G. McCrum, B.A. Read, G. Williams, "Anelastic and Dielectric Properties of Polymers", Wiley: New York (1967).
172. G.P. Johari, Ann. N.Y. Acad. Sci., 279, 117 (1976).
173. E.A. Guggenheim, Trans Faraday Soc., 45, 714 (1949).
174. E.A. Guggenheim, Trans Faraday Soc., 47, 573 (1951).
175. J.W. Smith, E.A. Guggenheim, Trans Faraday Soc., 46, 394 (1950).
176. E. Riande, E. Saiz, "Dipole, moments and Birrefrindgence of Polymers", Prentice Hall: Englewood Cliffs, NJ (1992).
177. MOPAC, Quantum Chemistry Exchange Program, Department of Chemistry Indiana University, Bloomington, IN.
178. Tripos Associates Inc., St. Louis, MO 63144.
179. M. Clark, R.D.III Cramer, N.J. van Opdembosch, Comput. Chem., 10, 983 (1989).
180. L. Onsager, J. Am. Chem. Soc., 58, 1486 (1936).
181. J. Kirkwood, J. Chem. Phys., 7, 911 (1939).
182. B.E. Tate, Polymer Sci., 5, 214 (1967).
183. D. Radić, L. Gargallo, Polym. Mater. Encyclopedia, 8, 6346, (1996).
184. A. Leon, M. López, L.Gargallo, D. Radić, A. Horta, J. Macromol. Sci-Phys, B, 29, 351 (1990).
185. J.M.G. Cowie, Z. Haq, Br. Polym. J., 9, 241 (1977)
186. J. Velicković, S. Vasović, Makromol. Chem., 153, 207 (1972).
187. J. Velicković, S. Coseva, R. Fort, Eur. Polym. J., 11, 377 (1975).
188. J. Velicković, M. Plavsić, Eur. Polym. J., 19, 1171 (1983).
189. J. Velicković, J. Filipović, Polym. Bull., 5, 569 (1981).
190. J.M.G. Cowie, I.J. McEwen, Eur. Polym. J., 18, 555 (1982).
191. L. Gargallo, D. Radić, A. León, Makromol. Chem., 186, 1289 (1985).
192. A. León, L. Gargallo, A. Horta, D. Radić, J. Polym. Sci. Polym. Phys., Ed., 27, 2337 (1989).
193. A. Ribes Greus, R. Díaz Calleja, L. Gargallo, D. Radić, Polymer, 32, 2755 (1991).
194. R. Díaz-Calleja, L. Gargallo, D. Radić, Polymer, 33, 1406 (1992).
195. R. Díaz-Calleja, L. Gargallo, D. Radić, J. Appl. Polym. Sci., 46, 393 (1992).
196. L. Gargallo, D. Radić, D. Bruce, Polymer, 34, 4773 (1993).
197. J.M.G. Cowie, R. Ferguson, J. Polym. Sci., Polym. Phys., Ed., 23, 2181 (1985).
198. A.F. Miles, J.M.G. Cowie, Eur. Polym. J., 27, 165 (1991).
199. J.M.G. Cowie, Z. Haq. Br. Polym. J., 9, 246 (1977).
200. J.M.G. Cowie, I.J. McEwen, M.Y. Pedram, Macromolecules, 16, 1151 (1983).
201. J.M.G. Cowie, R. Ferguson, I.J. McEwen, M.Y. Pedram, Macromolecules 16, 1155 (1983).
202. D. Radić, L.H. Tagle, A. Opazo, A. Godoy, L. Gargallo, J. Thermal Analy., 41, 1007 (1994).
203. D. Radić, A. Opazo, L.H. Tagle, L. Gargallo, Thermochimica Acta, 211, 255 (1992).
204. A. León, L. Gargallo, A. Horta, D. Radić, J. Polym. Sci., Polym. Phys., Ed., 27, 2337 (1989).
205. A. Ribes Greus, R. Díaz Calleja, L. Gargallo, D. Radić, Polymer, 32, 2755 (1991).
206. D. Radić, L. Gargallo, "Polymeric Materials Encyclopedia", New York, London, Tokyo: CRC Press, Boca Raton, 8, 6346 (1996).
207. A. León, L. Gargallo, A. Horta, D. Radić, J. Polym. Sci. Polym. Phys., Ed., 27, 2337 (1989).
208. A. Ribes-Greus, R. Díaz calleja, L. Gargallo, D. Radić, Polymer, 32, 2331 (1991).
209. R. Díaz-Calleja, L. Gargallo, D. Radić, Polymer, 33, 1406 (1992).
210. R. Díaz-Calleja, L. Gargallo, D. Radić, J. Appl. Polym. Sci., 46, 393 (1992).
211. R. Díaz-Calleja, E. Saiz, E. Riande, L. Gargallo, D. Radić, Macromolecules 26, 3795 (1993).
212. R. Díaz-Calleja, E. Saiz, E. Riande, L. Gargallo, D. Radić, J. Polym. Sci., Phys. Ed., 32, 1069, (1994).
213. J.M.G. Cowie, "Order in the Amorphous State of Polymers", New York: Plenum (1987).
214. J.M.G. Cowie, Pure. Appl. Chem., 51, 2331, (1979).
215. J.M.G. Cowie, J. Macromol. Chem. Sci. Phys., B18, 563 (1980).

216. A. Pérez-Dorado, I. Fernández-Piérola, J. Baselga, L. Gargallo, D. Radić, Makromol. Chem., 190, 2975, (1989).
217. A. Pérez-Dorado, I. Fernández-Piérola, J. Baselga, L. Gargallo, D. Radić, Makromol. Chem., 191, 2905, (1990).
218. I. Hernández-Fuentes, A. Horta, L. Gargallo, C. Abradelo, M. Yazdani-Pedram, D. Radić, J. Phys. Chem., 92, 2974 (1998).
219. E. Saiz, A. Horta, I. Hernández-Fuentes, L. Gargallo, C. Abradelo, D. Radić, Macromolecules, 21, 1736 (1988).
220. A. Horta, I. Hernández-Fuentes, L. Gargallo, D. Radić, Makromol. Chem. Rapid. Commun, 8, 523 (1987).
221. J. Velickvić, J. Filipović, Makromol. Chem., 185, 569 (1989).
222. J. Velicković, J. Filipović, S. Coseva, Eur. Polym. J., 15, 521 (1979).
223. M. Yazdani-Pedram, L. Gargallo, Eur. Polym. J., 21, 461 (1985).
224. M. Yazdani-Pedram, L. Gargallo, Eur. Polym. J., 21, 707 (1985).
225. L. Gargallo, M. Yazdani-Pedram, D. Radić, A. Horta, Eur. Polym. J., 25, 1059 (1989).
226. D. Radić, C. Dañin, A. Opazo, L. Gargallo, Makromol. Chem., Macromol. Symp., 58, 209 (1992).
227. L. Gargallo, M. Yazdani-Pedram, D. Radić, A. Horta, J. Bravo, Eur. Polym. J., 29, 609 (1993).
228. L. Gargallo, D. Radić, J. Macromol. Sci-Phys., B33, 75 (1994).
229. J.M.G. Cowie, Z. Haq. Polym., 19, 1052 (1978).
230. J.M.G. Cowie, N. Wadi, Polymer, 26, 1566 (1985).
231. J.M.G. Cowie, N. Wadi, Polymer, 26, 1571 (1985).
232. J.M.G. Cowie, A.C.S. Martin, Polym. Commun., 26, 298 (1985).
233. J.M.G. Cowie, Polymer, 28, 627 (1987).
234. S. Nagai, K. Yoshida, Bull. Chem. Soc. Jpn., 38, 1402 (1985).
235. R. Ottenbrite, K. Kruus, A. Kaplan, Polym. Prep., 24, 25 (1983).
236. E. Hodnett, J. Amirmoazzami, J. Tai, J. Med. Chem., 21, 652 (1978).
237. R. Díaz-Calleja, M.J. Sanchis, L. Gargallo, D. Radić, J. Polym. Sci., Polym. Phys. Ed. 34, 2616 (1996).
238. R. Díaz-Calleja, L. Gargallo, D. Radić, Macromolecules, 28, 6963 (1995).
239. J.M.G. Cowie, S.A. Henshall, I.J. McEwen, J. Velicković, Polymer, 18, 612 (1977).
240. J.M.G. Cowie, I.J. McEwen, J. Velicković, Polymer, 16, 664 (1975).
241. J.M.G. Cowie, I.J. McEwen, J. Velicković, Polymer, 16, 689 (1975).
242. E.A.W. Hoff, D.W. Robinson, A.H. Willbourn, J. Polym. Sci. Polym. Phys., 18, 161 (1955).
243. E. Sanchez, R. Díaz Calleja, P. Berges, J. Kudning, G. Klar, Synth. Met., 32, 79 (1989).
244. E. Sanchez, R. Díaz-Calleja, P. Berges, W. Gunsser, G. Klar, Synth. Met., 30, 67 (1989).
245. A Ribes Greus, R. Díaz-Calleja, L. Gargallo, D. Radić, Polymer 32, 2755 (1991).
246. R. Díaz-Calleja, L. Gargallo, D. Radić Polym. Int. 29, 159 (1992).
247. R. Díaz-Calleja, L. Gargallo, D. Radić J. Polym. Sci., Polym. Phys. Lett. 31, 107 (1993).
248. M. Yazdani-Pedram, E. Soto, L.H. Tagle, F.R. Díaz, L.Gargallo, D. Radić, Thermochim. Acta 105, 149 (1986).
249. L.H. Tagle, F.R. Díaz, L. Rivera, Themochim. Acta, 118, 111 (1987).
250. L.Gargallo, E. Soto, F.R. Díaz, L.H. Tagle, D. Radić, Eur. Polym. J., 23, 571 (1987).
251. M.J. Fabre, L.H. Tagle, L.Gargallo, D. Radić, I. Hernández-Fuentes, Eur. Polym. J., 25, 1315 (1989).
252. L.Gargallo, L.H. Tagle, D. Radić, Br. Polym. J., 22, 27 (1990).
253. E. Saiz, M.J. Fabre, L. Gargallo, D. Radić, I. Hernández-Fuentes, Macromolecules, 22, 3660 (1989).
254. E. Saiz, C. Abradelo, J. Mogin, L.H. Tagle, M.J. Fabre, I. Hernández-Fuentes, Macromolecules, 24, 5594 (1991).
255. I. Hernández-Fuentes, F. trey- Stolle, L.H. Tagle, E. Saiz, Macromolecules, 25, 3294 (1992).
256. M.J. Fabre, I. Hernández-Fuentes, L. Gargallo, D. Radić, Polymer, 33, 134 (1993).

257. E. Sanchez-Martinez, R. Díaz Calleja, S. Monserrat Ribas, L. Gargallo, D. Radić, Polym.
      Int., 28, 227 (1992).
258. P.R. Sundararajan, Macromolecules, 22, 2149 (1989).
259. P.R. Sundararajan, Macromolecules, 23, 2600 (1990).
260. P.R. Sundararajan, Macromolecules, 26, 344 (1993).
261. J. Bicerano, H.A. Clark, Macromolecules, 21, 585 (1988).
262. J. Bicerano, H.A. Clark, Macromolecules, 21, 597 (1988).
263. M. Hutnik, A.S. Argon, U.W. Suter, Macromolecules, 24, 5956 (1991).
264. PT-17. M. Hutnik, A.S. Argon, U.W. Suter, Macromolecules 24, 5970 (1991).
265. R. Díaz Calleja, D. Radić, L. Gargallo J. Non-Cryst. Solids, 172, 907 (1994).

Chapter 3
Physicochemical Aspects of Polymer at Interfaces

Summary A discussion of the behavior of polymers at interfaces where the Langmuir monolayers and Langmuir–Blodgett films are studied. Amphiphilic polymers at the air–water interface are studied through the Langmuir technique. The study and discussion of surface pressure-area isotherms for different polymers are performed by using a surface film balance and the results obtained from this technique are analyzed in terms of the shape of the isotherms. The collapse pressure for different systems is discussed in terms of the chemical structure of the polymer. The adsorption of polymers by spreading and from solution is also discussed. Wetting of solids by a liquid described in terms of the equilibrium contact angle θ. At the appropriate interfacial tensions in the equilibrium the forces acting are analyzed using the Young's equation.

Keywords Interface · Langmuir monolayer · Adsorption process · Spreading solvent · Surface tension · Surface isotherm · Langmuir–Blodgett film · Collapse pressure · Amphiphilic polymer

3.1 Introduction

The forces acting between two surfaces in contact or near – contact determine the behavior of a wide spectrum of physical properties. These can include friction, lubrication, the flow properties of particulate dispersions, and, in particular, the adsorption and adhesion phenomena, the stability of colloidal system [1, 2] and the ability to form Langmuir monolayer at the air – water interface.

Polymers at interfaces are an important part of a large number of polymeric technological applications. The orientation, specific interactions, and higher order structures of amphiphilic molecules at a quasi two- dimensional plane form the basis of a variety of interesting phenomena [3–6]. Scheme 3.1 shows a schematic representation of this behavior.

The location of a molecule at the interface between two immiscible fluids may lead to a variety of organizational and dynamic interfacial behavior dependent on the relatively of the components of the chemical structure of the molecule in each of the

L. Gargallo, D. Radić, *Physicochemical Behavior and Supramolecular Organization of Polymers*, DOI 10.1007/978-1-4020-9372-2_3,
© Springer Science+Business Media B.V. 2009

(a) short amphiphilic molecules

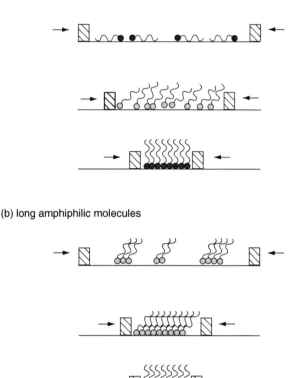

(b) long amphiphilic molecules

Scheme 3.1 Shows the squematic representation of the orientation of one short and long amphiphilic molecule under different conditions. (From ref. [5])

phases, two solvents or one solvent and air. In the extreme case, the molecule could detach completely from the interface and disperse into one or both liquid phases as individual molecule or aggregates. When the molecule is a macromolecule or a polymer, the interactions between polymers adsorbed onto surface are important in many processes.

For example, it is believed that polysaccharides incorporated into the membrane of a cell are responsible for the cell adhering onto a surface, and, in different area, polymers adsorbed onto surfaces play an important role in stabilizing industrial colloidal dispersions such as paints and inks.

It is from a "colloid" point of view that experiments have been performed to understand the behavior of polymers at interfaces [7].

There are very important phenomena originated when a polymer is on one interface or close to one interface. Such phenomena include adsorption, adhesion, monolayer formation, coating and colloidal stabilization. In the majority of circumstances a polymer adsorbed onto colloidal particles will increase the stability of a dispersion [8]. The essential feature of polymer or "steric" stabilization is

Fig. 3.1 Schematic
representation for steric
stabilization. (From ref. [8])

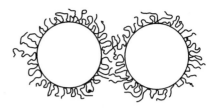

that the adsorbed polymer keeps the core of the particles sufficiently far apart so
that the attractive van der Waals' interactions between the particles are not strong
enough to cause aggregation. Figure 3.1 shows a schematic representation for a
steric stabilization.

A polymer molecule may also simultaneously adsorb onto both surface, hence
causing flocculation by polymer-bridging mechanism [8]. Figure 3.2 shown this
behavior.

Many other applications of polymers involve their use in multiple phases, such as
composites, reinforced rubbers and adhesives. Consequently, the interaction of the
polymer and the other phase plays an important role in the physical properties of
the system. The properties of polymers at interfaces have been reviewed [1, 8–11].
A number of techniques have been used to study the structure of polymers at interfaces such as nuclear magnetic resonance (NMR).

Conventionally, the surface energetic properties of solids in general, are assumed
to be independent of the environment (polar and nonpolar) in which they are im-
mersed. However, Holly and Refojo [12], taking into account contact angle and
hysteresis experiments have concluded that hydrogel surfaces changed their polarity
according to the nature of the environment. Such surface modification phenomena
have been investigated by employing the ESCA method [13, 14], surface acidity
titration [15], contact angle, hysteresis [12, 16] and dynamic contact angle measurements [17, 18]. Environmentally induced changes in the surface structure have been
observed in those polymers which have relatively mobile surface layers and contain
polar moieties. The polarity is in some cases incorporated by surface modification
techniques, such as plasma treatment [19, 20], chemical oxidation [21], grafting
hydrophilic polymers onto hydrophobic polymers [22–24]. Lee et al. [25] have followed the surface modification behavior of polymers of different hydrophilicities
and rigidities by dynamic contact angle measurements and they have evaluated the
time scale of their restructuring. Numerous methods of polymer surface modification have already been discussed in the literature [26–28]. One method which
have paid considerable attention in the literature is the modification of polymer

Fig. 3.2 Schematic
representation for the
mechanism of
polymer-bridging
flocculation. (From ref. [8])

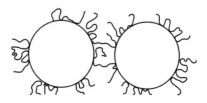

surfaces by plasma treatment. This allows the very rapid chemical modification of the top few monolayers of material. Two other common methods of polymer surface modification are ozone treatment [29] and corona discharge [30]. The surface and interface properties of polymers depend on both the processing conditions used, and the chemical compositions of the polymer. The surface properties can change in an unpredictable way due to contamination, weathering and migration of additives or groups. Alternatively, the surface can be deliberately changed by chemical reactions, electric discharges, plasma and ion beams and chemical or physical etching. The thickness of the surface layer affected by these changes will vary in significative form [31].

The question of polymers or and other polymers such proteins adsorb reversibly or not has been also a controversial subject. There has been a widespread belief that adsorption of polymers or proteins at the air – water interface is irreversible. This is based mainly on two observations. One is that protein molecules change from a globular shape in solution to an unfolded state at an interface. However, it is very well known from polymer chemistry that large molecules change their configuration when they move from one energy environment to another. The surface is a very different environment from the bulk solution. The change to an extended configuration in which the protein molecule lowers the free energy of the system by maximizing contact of nonpolar side chains with the nonpolar phase, within steric limitations, is understandable [32].

The other observation is that polymers or proteins, although highly soluble in aqueous solution, are found to be insoluble at the surface in the sense that compression to surface pressures of about $20\,mNm^{-1}$ does not usually result in detectable losses of monolayer. Despite this, desorption measurements have been reported by a number of workers [33–35]. Some results show that the desorption of proteins can be measured under suitable conditions and does not support the view that protein adsorption is an irreversible process. In general, as the molecular weight of the protein increases, the rate of desorption at a given surface pressure decreases markedly [7]. The conclusion is that the polymers or proteins of higher molecular weight require high surface pressures ($>30\,mNm^{-1}$) for significant rates to be measured. Thus observations on macromolecules which showed little or no desorption when compressed to pressures of $20–25\,mNm^{-1}$, although valid, do not necessarily signify irreversibility.

With the increasing interest in ultrathin organic films for applications as electronic and electrooptical devices and as models for biological membranes, the Langmuir – Blodgett (LB) technique has received considerable attention [36–38]. One of the advantages of using the Langmuir – Blodgett technique (LB) is that films of desired structures with thicknesses controlled at the molecular level can be obtained. Initially, research primarily focused on mono and multilayers of low molecular weight molecules. However, these films suffer from thermal and mechanical instability and are sensitive toward environmental attack. Because to their superior stability, LB films of polymers are nowadays studied for their application in optical and electronic devices.

3.2 Langmuir Monolayers and Langmuir-Blodgett Films

Studies on polymer monolayers spread at air – water interface to characterize their physical properties, as surface pressure, π, surface potential, ΔV, surface viscosity, η_S, and surface rigidity [39, 40] have been reported abundantly.

These properties have been measured as a function of the surface area, A, which can be expressed as square Angstroms per monomer unit, Å or in many cases, for convenience, by square meters per milligram of a polymer, Γ. However, over the last two decades a great deal of data have been accumulated on the monolayer formation (Langmuir monolayer) of preformed polymers on water and their spreading behavior [41–43].

Polymer monolayers on the aqueous subphase are studied using the Langmuir technique as is shown in Scheme 3.2.

Surface pressure-area (π – A) isotherms are obtained normally using a surface film balance. In those works, extrapolated area and collapse pressure determined from the surface pressure – area curves have been usually interpreted and compared with projected area of a monomeric unit by taking into account of chemical structure of the polymers.

Figure 3.3 shows as an example, the surface pressure-area per repeating unit (π – A) isotherms of Langmuir monolayers obtaining by spreading of poly(N-vinyl-2-pyrrolidone) (PVP) on aqueous Na_2SO_4 subphase, for both molecular weights and several temperatures [44].

For both high and low molecular weights, there is a temperature dependence. It is also clear from these Figures that the critical surface pressure values (π_C) depend markedly on temperature. The collapse state is characterized by the collapse pressure or critical surface pressure (π_C). This surface pressure can be defined as the maximum pressure value that the monolayer can reach without expulsion or rejection of the material in order to form a new tridimensional phase.

The collapse or critical pressure frequently is a highly controversial subject, but in this collapse area, the compression of the monolayer is not possible without destroying the monolayer.

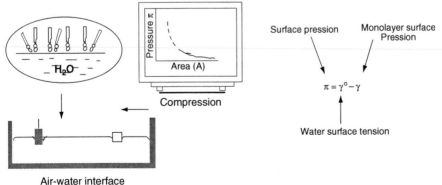

Scheme 3.2 Spreading monolayer Langmuir technique

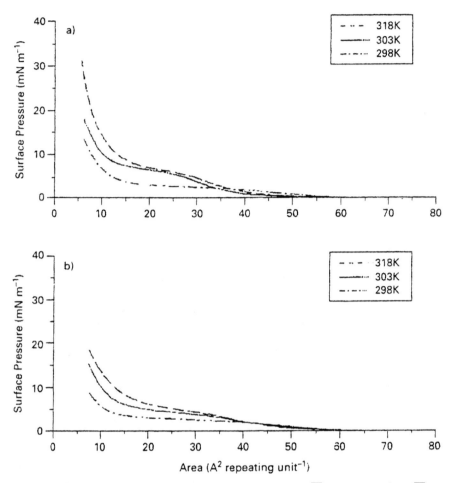

Fig. 3.3 Surface pressure-area ($\pi - A$) isotherms for PVP with (**a**) $\overline{M}_W = 3.65 \times 10^5$; (**b**) $\overline{M}_W = 0.42 \times 10^5$ on a 0.55 M aqueous Na_2SO_4 solution subphase at three temperatures: 298, 303, and 318 K. (From ref. [44])

In Fig. 3.3a and b, it was possible to observe the maximum value of π_C at the different temperatures that the monolayer can reach. At higher temperatures, π_C increases, PVP exhibits a low critical solution temperature (LCST) in water [45]. The PVP in 0.55 M aqueous Na_2SO_4 exhibits a lower LCST at 301 K. For soluble amphiphilic monolayers, such as those of PVP, when the temperature is changed the loss of monolayer material must be considered due to the solubilization in the subphase.

Different molecular areas of Langmuir monolayers can be determined. They can be defined in three ways: A_0 is the area per molecule extrapolated to zero differential surface tension, A_c is the minimum area per molecule at the collapse point, at the point in the $\pi - A$ isotherms where the pressure is the maximum reversible pressure (or "collapse pressure" π_c); and A_m is the area at the midpoint pressure $\pi_m = 0.5\,\pi_c$.

Table 3.1 Monolayer data from pressure-area $(\pi - A)$ isotherms for I_a, I_b, II_a, II_b and II_c on water subphase at 298 K. (From ref. [46])

		on water subphase at 298 K			
Polymer	π_c (mNm^{-1})	A_c, Å^2 experimental values	A_m, Å^2	A_o, Å^2	A_c, Å^2 estimated values by MDS
I_a	53	6.5	15	23	6.4
I_b	54	8.1	17	24	9.2
II_a	47	10	25	33	11.1
II_b	47	12.5	32	41	12.2
II_c	55	7	18	27	6.1

Table 3.1 shows, as an example, the different molecular areas values that has been reported for some poly(ester)s containing Si and Ge in the main chain [46].

The chemical structure of this type of poly(ester)s are shown in Schemes 3.3 and 3.4.

The structure and molecular orientation of the monolayers have been studied taking in account the different conformations of the repeating unit in the collapse area. Molecular Dynamic Simulation (MDS) has been also reported. The MDS un-

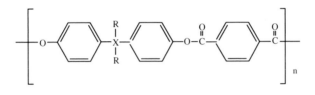

$$I_a : R = -C_6H_5; \; X = Si$$
$$I_b : R = -C_6H_5; \; X = Ge$$

Scheme 3.3 Poly(ester)s derived from diphenols containing silicon (Si) or germanium (Ge) and terephtalic acid dichlorides. (from ref. [46]).

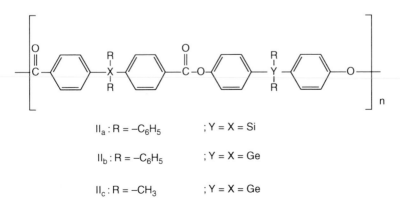

$$II_a : R = -C_6H_5 \qquad ; Y = X = Si$$

$$II_b : R = -C_6H_5 \qquad ; Y = X = Ge$$

$$II_c : R = -CH_3 \qquad ; Y = X = Ge$$

Scheme 3.4 Poly(ester)s containing two heteroatoms, Si or Ge in the main chain. (From ref. [46]).

der periodic condition was made in order to describe the experimental area values observed with the results shown in Table 3.1. The areas per molecule, A_c given in Table 3.1 are consistent with the minimum size of the repeating unit of the poly(ester)s estimated by MDS [46].

From the calculation of MDS it was possible to conclude that the polymers exist in the zigzag form in the monolayer with the ester groups toward the water and phenyl part toward the air. This organization can explain the largest areas for the polymers IIa and IIb as summarized in Table 3.1. To calculate the interaction energy of Ia, Ib, IIa, IIb and IIc, two conformations of each polymer segment fragment were selected randomly from the data collection. These are initially placed so that their center of mass coincides. One polymer fragment is then oriented at a random angle relative to the polymer fragment and the center of mass the second fragment is translated along the a certain vector until it meets the van der Waals surface of the first. This process was repeated to generate 100,000 polymer fragment pairs. Then the interaction energy was calculated for each polymer fragment pair as it can observed in Fig. 3.4.

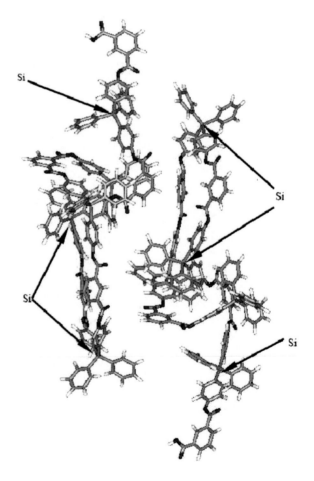

Fig. 3.4 Molecular models of the polymer-polymer interaction for poly(ester)s containing two Si atoms. (From ref. [46])

Table 3.2 Coulombic, van der Waals, and total interaction energy and data from molecular mechanics study. All the energies are in units of kcal/mol. (From ref. [46])

Polymers	$E_{(Coulombic)}$	$E_{(vdw)}$	$E_{(total)}$
I_a	−5, 3	−22, 5	−21, 6
I_b	−3, 1	−24, 3	−24, 9
II_a	−5, 0	−22, 0	−21, 2
II_b	−4, 5	−19, 6	−20, 5
II_c	−4, 3	−21, 1	−20, 8

Table 3.2 shows the additive total interaction energy, the Coulombic, and the van der Waals interaction energy values from the molecular mechanics study for each polymer fragment.

The energy contribution from the MDS study was found mainly from the van der Waals interaction, being less favorable for the poly (ester)s with one Si (Ia) than with one Ge atom (Ib). These results have been in good agreement with the dispersion contribution to the total surface energy which was estimated by wettability measurements [46].

As it has been remarked studies on polymer monolayers spread at the air – water interface to characterize their surface behavior as a function of area, of

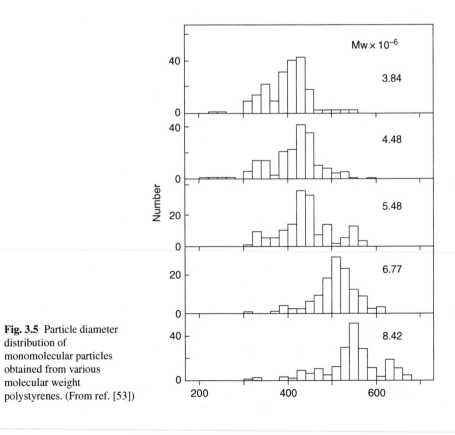

Fig. 3.5 Particle diameter distribution of monomolecular particles obtained from various molecular weight polystyrenes. (From ref. [53])

many polymers are very abundant [47–52] But, evidently, not all synthetic polymers can form stable Langmuir monolayers. Fully nonpolar hydrocarbon – based polymers such as poly(ethylene) and poly(styrene) cannot be spread on water because of the lack attractive interaction with a water surface. In the particular case of poly(styrene), Kumaki [53] has reported results about the obtention of ultrafine particles, each containing a single poly(styrene) molecule by spreading dilute solutions of poly(styrene) in benzene onto a water surface. For convenience these particles were called "monomolecular particles", but not monolayer.

Figure 3.5 shows the particle diameter distributions for the particles corresponding to the observation by transmission electron microscopy. From these results a clear molecular weight dependence of diameter distribution can be observed.

A polymer with different hydrophilic and hydrophobic parts are necessary for monolayer formation on a water subphase, this corresponds to an amphiphilic polymer, a type of polymers frequently used to prepare mono- and multilayers [54, 55]. This situation can be illustrated in Fig. 3.6, for a short molecule as stearic acid.

Some polymers with different hydrophobic and hydrophilic parts are summarized in Scheme 3.5 (From ref. [23]).

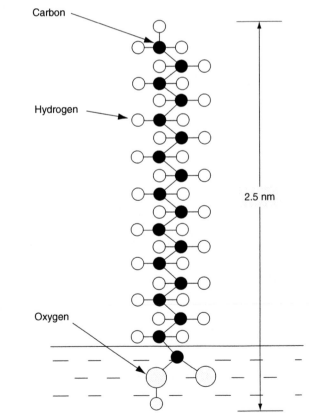

Carbon

Hydrogen

2.5 nm

Oxygen

Fig. 3.6 Molecular structure of stearic acid. (From ref. [56])

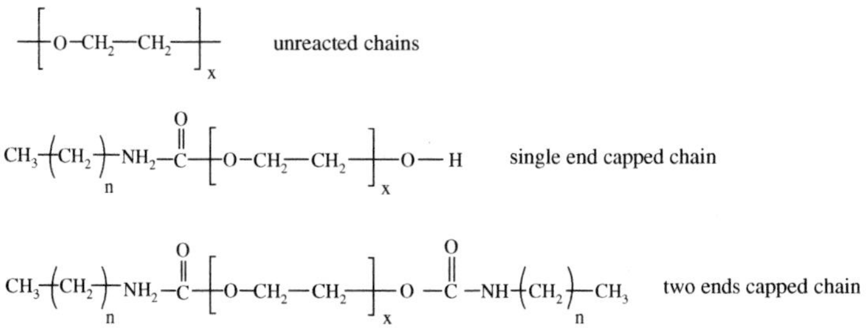

$$\left[O-CH_2-CH_2 \right]_x \qquad \text{unreacted chains}$$

$$CH_3\left(CH_2\right)_n-NH_2-\overset{\overset{O}{\parallel}}{C}\left[O-CH_2-CH_2\right]_x O-H \qquad \text{single end capped chain}$$

$$CH_3\left(CH_2\right)_n-NH_2-\overset{\overset{O}{\parallel}}{C}\left[O-CH_2-CH_2\right]_x O-\overset{\overset{O}{\parallel}}{C}-NH\left(CH_2\right)_n-CH_3 \qquad \text{two ends capped chain}$$

X = 23	PEO1-Octyl	n = 8
	PEO1-Dodecyl	n = 12
	PEO1-Octadecyl	n = 18
X = 203	PEO2-Octyl	n = 8
	PEO2-Dodecyl	n = 12
	PEO2-Octadecyl	n = 18

Scheme 3.5 Poly(ethylene oxide)s of different molecular weight hydrophobically modified with alkyl isocyanates of different length chain. (Fron ref. [23]).

This scheme shows the possible chemical structures obtained for poly (ethylene oxide)s hydrophobically modified.

Poly(ethylene oxide)s of different molecular weight hydrophobically modified with alkyl isocyanates of different length chain.

The surface isotherms obtained with these hydrophobic modified polymers were interpreted in terms of hydrophobic and hydrophilic balance of the polymers.

From the surface pressure – area isotherms of these polymers, the surface parameters could be calculated. The areas per monomer unit projected to zero surface – pressure, obtained from the linear variation of π with the surface concentration (A_0) in semidilute region [39] for the polymers, are summarized in Table 3.3.

The values obtained in all cases were very similar. Figure 3.7 shows this behavior. But, according to Leiva et al. [23] the values occupied per repeat unit in adsorbed monolayer (σ) were larger than those of the spread monolayer.

Table 3.3 Area for monomer unit projected to zero surface pressure (A_0) for PEO 1000 and 10.000 g/mol and their derivatives. (From ref. [23])

Polymer	$A_0(\text{Å}^2/\text{m.n.})$
PEO1	23 ± 2
PEO1-octyl	22 ± 2
PEO1-dodecyl	21 ± 2
PEO1-octadecyl	22 ± 2
PEO2	32 ± 2
PEO2-octyl	30 ± 2
PEO2-dodecyl	35 ± 2
PEO2-octadecyl	33 ± 2

Fig. 3.7 Show the Lanmuir isotherms where the surface concentration was expressed as area per repeat unit ($\mathring{A}_2$/r.u.) and show graphic determination of A_0 for PEO2 and PEO1-octadecyl. (From ref. [23])

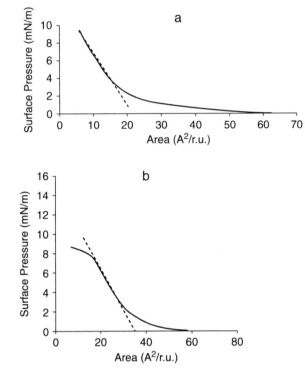

Table 3.4 summarizes the adsorption parameters obtained from adsorption isotherms of PEO and their derivatives. Γ^∞ and σ are the limiting excess surface concentration and the area covered per monomer unit respectively. In the same Table are also summarized the pC_{20} and ΔG^0 values. pC_{20} is a convenient estimation of the efficiency of adsorption and is defined as the negative logarithm of the concentration of surfactant in the bulk phase required to produce a $20\,mNm^{-1}$ reduction in the surface or interfacial tension of the solvent ($-\log C(\Delta\,\gamma = 20) = pC_{20}$). ΔG^0_{ads} is the classical free energy change on adsorption at infinite dilution.

It is very interesting the adsorption behavior of poly (ethylene oxide) from water solution to the air – water interface. This adsorption was considerably enhanced by a hydrophobic group placed at ends of the molecules.

Figure 3.8 shows the squematic representation of this adsorption process.

Table 3.4 Adsorption parameters obtained from adsorption isotherms of the PEO 1000 g/mol, their derivatives, and PEO 10,000 g/mol (From ref. [23])

Polymer	Γ^∞ (mol/m²)	σ (m²/m.u.)	σ ($\mathring{A}^2$/r.u.)	pC_{20}	ΔG^0_{ads} (kJ/mol)
PEO1	1.28×10^{-6}	1.30×10^{-18}	124.7	–	–
PEO1-octyl	4.96×10^{-6}	3.35×10^{-19}	33.5	2.2	−26.5
PEO1-dodecyl	4.19×10^{-6}	3.96×10^{-19}	39.6	2.2	−27.2
PEO1-octadecyl	4.54×10^{-6}	3.66×10^{-19}	36.6	2.0	−25.7
PEO2	1.52×10^{-6}	1.10×10^{-18}	109.5	–	–

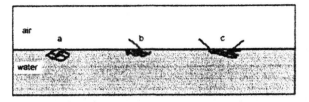

Fig. 3.8 Schematic representation of (**a**) PEO chain adsorbed at the air-water interface. (**b**) single end capped PEO chain adsorbed at the air-water interface, and (**c**) double end capped chain adsorbed at the air-water interface. (From ref. [23])

To analyze the surface activity of one component, the minimum surface tension (γ_{min}) can be calculated as:

$$\gamma_{min} = \gamma_s - \pi_{max} \tag{3.1}$$

where here, π_{max} is the maximum surface pressures and γ_s is the surface tension of the subphase. Relative limiting area (A_o) and relative lift off area (A_1) are quantities which indicate molecular packing and interaction of molecules in the monolayer.

The relative limiting area is determined by extrapoling the final steep linear region of the isotherm at end compression to the % area axis. The relative lift off area is obtained by extrapolating the area at which an increase in surface pressure from the baseline value is observed to the % area axis.

Another surface parameter of interest is the hysteresis area (ΔG), which is indicative of energy trapped in a monolayer. The hysteresis area is the difference between the free energy of compression and free energy of expansion which is calculated from the area under corresponding surface pressure – area isotherms.

The two dimensional compressibility (C_s) of the monolayer is calculated as

$$C_s = (1/A)(dA/d\pi), \tag{3.2}$$

Where $d\pi/dA$ is the slope of the surface pressure – area isotherm and A is the relative area of the film. C_s will decrease with the amount of surface active material present and will be less for more condensed films [57].

By comparing the extrapolated area to zero surface – pressure from the surface pressure – area ($\pi - A$) isotherms with projected area of a monomer unit from molecular models of the polymer segments, it is possible to obtain information on molecular orientation and packing of these polymer monolayers.

However, as Ringard-Lefebvre et al. [58] have noted, such comparisons are only valid if the assumption that polymer molecules which in bulk solution form tightly bound three-dimensional coils could, during the spreading process from organic solvents onto aqueous subphase, extend to cover the available area with every polymer segment on surface layer.

In the polymer monolayer spread at the air – water interface, the surface concentration can be regulated easily by compression or expansion of the monolayer. It is

Fig. 3.9 Static
compressibility modulus
$\varepsilon = d\pi/d \ln \Gamma$, at 25°C,
calculated from the π vs Γ,
curve. The dashed lines
indicate the limits of the
dilute (D), semidilute (SD),
and concentrated regimes
(C). (From ref. [59])

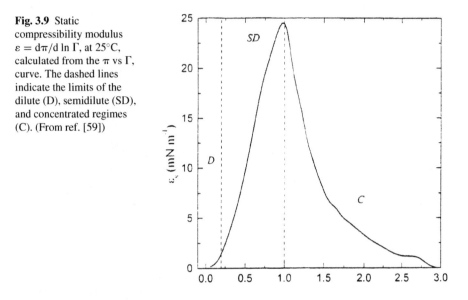

possible to make an analogy between the intervals of superficial concentration and
the regimes of concentrations of solutions frequently used in polymeric science.

Figure 3.9 shows a squematic representation of the different regimen of surface
concentrations.

By this analogy, the extrapolated area to zero surface pressure and the collapse
pressure responds to phenomena that happen in the concentrated region (in two
dimensions) [59, 60]. In the semidilute concentration region, surface pressure obeys
to power law of the surface concentration and is independent on the molecular
weight. In this concentration regime, that is, in moderately concentrated solution,
the polymer chain partially interpenetrate each other. According to the scaling concepts, [61,62] the surface pressure in the semidilute region varies with the superficial
concentration according to the follow expression [63]:

$$\pi = \Gamma^{2\nu/(2\nu-1)} \tag{3.3}$$

where Γ is the surface concentration in mg/m^2 and ν is the critical exponent of the
excluded volume.

At low concentration region, colligative properties such as surface pressure are
best described by a virial expansion of the surface concentration Γ [59, 64].

$$\pi/(\Gamma RT) = 1/M_n + A_{2,2}\,\Gamma + A_{2,3}\,\Gamma^2 + \cdots \tag{3.4}$$

where R is the gas constant, T the absolute temperature, M_n the number average
molecular weight and $A_{2,2}$ and $A_{2,3}$ are the second and third virial coefficients at
two – dimensional space, respectively.

Leiva et al. [65] have reported for poly(itaconates) monolayers the surface behavior at the air – water interface at different surface concentrations. They have found that for these type of polymers, the air – water interface at 298 K, is a bad solvent, very close to the theta solvent. At the semidilute region concentration, the surface pressure variation was expressed in terms of the scaling laws as a power function of the surface concentration. According to equation (3.3), the $\log \pi$ vs $\log \Gamma$ plot shows a linear variation with slope $2\upsilon/(2\upsilon\text{-}1)$.

Figure 3.10 shows this double logarithmic plot of π vs Γ for six polymers studied.

Table 3.5 summarizes the ν values determined for the poly (itaconates) family: Poly(monooctyl itaconate) (PMOI); Poly(monodecyl itaconate) (PMDI); Poly (monododecyl iataconate) (PMDoI); Poly (methyldodecyl itaconate) (PMeDoI); Poly(benzyl itaconate)(PMBzI) and the copolymer Monooctyl -alt- Maleic anhydride (MOI-alt- MA).

For polymer chains in two dimensions in good solvents, the theoretical predictions point to a ν value narrowly centered in 0.75 [61]. Monte Carlo simulations predicts a value of 0.753 [66], while by the matrix-transfer method a value of 0.7503 is predicted. In the case of theta condition the situation is not clear, the predictions are less precise. Monte Carlo simulation [66] has suggested $\nu_\theta \approx 0.51$ while matrix-transfer data suggest $\nu_\theta \approx 0.55$ [67].

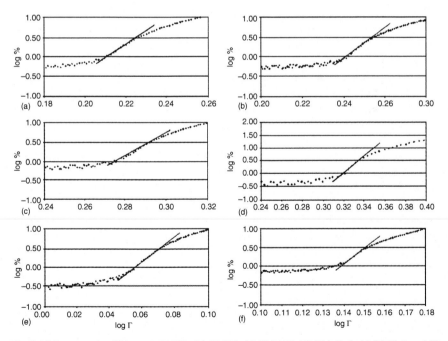

Fig. 3.10 $\log \pi$ vs $\log \Gamma$ for: (**a**) PMOI, (**b**) PMDI, (**c**) PMDoI, (**d**) PMeDoI, (**e**) PMBzI and (**f**) MOI-alt-MA. (From ref. [65])

Table 3.5 v values obtained from the linear region of Log π log Γ plots shown in Fig. 3.10 (From ref. [65])

Polymer	$v \pm 0.01$
PMOI	0.51
PMDI	0.51
PMDoI	0.52
PMeDoI	0.51
PMBzI	0.52
MOI-alt-MA	0.51

As it is possible to observe in Table 3.5, the values reported are very close to 0.51, which indicates proximity to the theta condition at the air – water interface for all these poly(itaconates) derivatives.

It is very well known that different macromolecular arrangements may be induced either by changing the nature of the subphase [44, 68, 69] or by changing the spreading solvent [70]. However, only a few studies describing these effects on polymer monolayer surface pressure behavior have been reported [71–73].

Maleev et al. [71] have been investigated the effect of different spreading solvents on compressional behavior of four polymer monolayers: Poly(methyl acrylate), poly(ethyl acrylate), poly (methyl methacrylate), poly (ethyl methacrylate). They have shown that the compressional behavior of poly (methacrylate) monolayers is affected by the nature of the solvent to a much higher degree than that of poly (acrylate) monolayers. They have related this behavior to the polymer glass transition temperature (Tg) and suggested that polymers having a Tg higher than the temperature at which their monolayers are spread are essentially independent of compression configuration, this is the case of poly(methacrylate) monolayers. Conversely, polymers which have Tg lower than the temperature at which their monolayers are spread may form several kinds of arrangements resulting from intramacromolecular chain rearrangements which occur on compression of these monolayers. This is the case of poly (acrylate).

Baglioni et al. [72] showed that poly (γ – methyl L – glutamate) may form a monolayer in the α helice or β sheet conformation according to whether the spreading solvent did or did not contain pyridine.

In general, there is little information available on how a change in spreading conditions can affect the behavior of polymer monolayers at the air – water interface.

Important differences in the contour of poly (D,L- lactic acid) π – A isotherms resulting from the type of organic solvent from which the polymer was spread over the water subphase require explanation [58]. The π – A isotherms of poly (D,L-lactic acid) monolayers spread from various solvents are shown in Fig. 3.11 [58].

The properties of poly(D, L-lactic acid) monolayers spread at the air – water interface were also shown to be strongly dependent of the nature of the spreading solvent. In this case, the monolayers spread from acetone and tetrahydrofuran exhibited typical reversible collapse behavior in the compression – expansion cycle with a quasi – plateau at large areas followed by a steep rise in the surface pressure at small areas. Conversely, the monolayers spread from chloroform, dichloromethane,

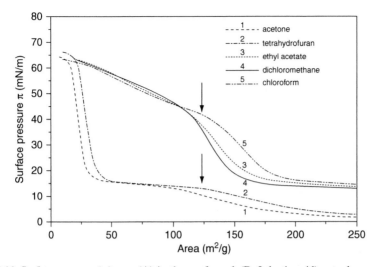

Fig. 3.11 Surface pressure (π)-area (A) isotherms for poly(D, L-lactic acid) monolayers spread at the air/water interface 250 μg of polymer was spread from (1) acetone, (2) tetrahydrofuran, (3) ethyl acetate, (4) dichloromethane, and (5) chloroform. Arrows indicate area (125 m^2/g) at which Langmuir-Blodgett sampling was performed. Compression rate 34.5 cm^2/min, (From ref. [58])

and ethyl acetate displayed large hysteresis characteristic of irreversible collapse and high surface pressures throughout the compression cycle [58].

Usually, the fabrication of a close – packed assembly of amphiphilic molecules at an air – water interface by the Langmuir method requires suitable subphase conditions related to the ionic species and its concentration, pH, temperature and addition of another complementary solutes. In the last case, to explore the feasibility of enhancing the interactions of some amphiphilic polymers with water soluble polymers at the air – water interface, it was studied the system of poly(monomethyl itaconate) (PMMeI) as subphase stabilizer of maleic anhydride – alt – stearyl methacrylate(MA-alt-StM) monolayers at the air – water interface.

Surface pressure [68] (π) vs. Area (A) (Å$_2$ per repeating unit) isotherms were measured for MA-alt-StM on water and on aqueous poly (monomethyl itaconate) solutions at pH 3.0 and 7.0. Figure 3.12 shows these isotherms.

It has been observed that PMMeI modified considerably the shape of the MA-alt-StM on water isotherms. The zero – pressure limiting area per repeating unit (ru), A$_0$, rose from 38 to 41 A$_2$/ru.

Table 3.6 summarizes the characteristic π – A isotherm curves of MA – alt – StM under different pH conditions.

At pH 3.0, the surface pressure along the MA – alt – StM isotherm on aqueous PMMeI as subphase was higher than that at pH 7.0 under otherwise identical conditions. At pH 3.0, PMMeI stabilized the MA – alt – StM monolayers, presumably via hydrogen – bonding which is believed to result in an interpolymer complex. The formation of the complex was supported in terms of Brewster Angle

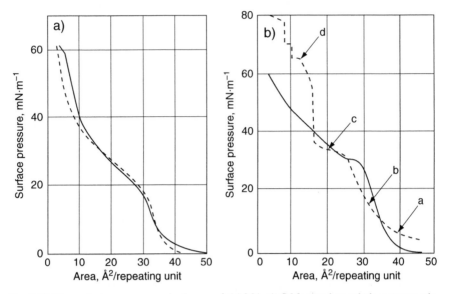

Fig. 3.12 The surface pressure-area isotherms of: (**a**) MA-alt-StM spread on subphase water solutions at pH 3.0 (---) and pH 7.0 (–); (**b**) MA-alt-StM on subphase water-PMMeI solution at pH 3.0 (---) at pH 7.0 (–); a, b, c and d are points at which the BAM images are shown. (From ref. [68])

Microscopy images, which showed the monolayer to be more homogeneous, and by FTIR spectra which adduced evidence for the occurrence of specific interactions.

Scheme 3.6 shows the chemical structure of the polymers studied. (From ref. [68]).

The results from surface – pressure – area (π – A) have been explained by the fact that the protonated carboxylic groups of PMMeI and the maleic anhydride groups can be able to form an interpolymer complex at the air – water interface. The FTIR spectra seem to indicate that the superior driving force for the complexation will be hydrogen – bonding between the proton – donating carboxyl group of PMMeI and the proton – accepting groups of the MA copolymer. The complex can also be stabilized by hydrophobic interactions. This last conclusion has been supported by the control experiment on surface behavior of PMMeI. The hydrophobicity of PMMeI in water increases as the pH was decreased [68].

In the past decades, extensive studies have been performed to elucidate the structures and intermolecular forces of two-dimensional arrays of molecules at the

Table 3.6 Characteristic π – A isotherm curves of MA-alt-StM under different pH conditions. (From ref. [68])

Water subphase	π_0, mN $\cdot$ m^{-1}	A_0, Å^2 $\cdot$ monomer^{-1}
PMMeI, pH 3.0	~ 70	41
PMMeI, pH 7.0	60	38
Pure water, pH 3.0	60	38
Pure water, pH 7.0	60	38

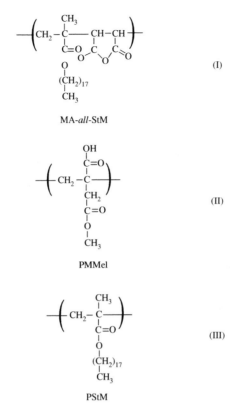

MA-*all*-StM

(I)

PMMeI

(II)

PStM

(III)

Scheme 3.6 Maleic anhydride-alt-stearyl methacrylate (MA-alt-StM), poly(monomethyl ita-conate) (PMMeI), poly(stearyl methacrylate) (PStM). (From ref. [68])

air-water interface. Particularly, there has been increasing interest in studying the basic properties of functionalized polymers at interfaces and the surface energy of the polymers [73–76]. The determination of the surface free energy of a polymer is of interest in different fields such as adhesion [77] and adsorption [78]. Interactions between polymer chains and interfaces (either liquid – solid or liquid – air) give rise to changes in the conformational and thermodynamic behavior of the chains relative to those in the bulk. The need for polymeric materials with amphiphilic properties for specific applications has led to the development of chemical modifications of some classical polymers [79].

In this context, poly(4-vinylpyridine)s quaternized with alkyl side – chains of different lengths and differents quaternization degrees have been prepared [79–81].

These synthetic hydrophilic polymers hydrophobically modified can be good systems with which to try to establish relationships between chemical structure and interfacial characteristics. It was determined the surface pressure – area (π – A) isotherms at the air – water interface for poly(4-vinylpyridine) quaternized as a function of the methylene group number of the alkyl lateral chains (n). The film formation of these polymers on aqueous subphase at constant pH and ionic strength has been studied by the Langmuir technique. The π – A isotherms of poly(4-vinylpyridine)s quaternized with four different alkyl chains (pentyl, hexyl, octyl and decyl bromide) (P4VPC$_5$Br), (P4VPC$_6$Br), (P4VPC$_8$Br) and (P4VPC$_{10}$Br), respectively, are shown in Fig. 3.13.

Depending on the length of the side-chain, the π – A isotherms show plateau regions. The difference in compression depends on the length of the side-chain. π – A isotherms of the octyl and decyl derivatives on pure water at pH 5.7 present very similar profiles (Fig. 3.13 curves c and d). These monolayers are of the condensed type [79]. The pentyl derivative of P4VP gives a different shape: a plateau region is not observed, and the collapse pressure is smaller (see Fig. 3.13 curve a). The quaternization of poly(4-vinylpyridine) has significant effects on its adsorption on aqueous surface. A critical methylene group number (CMG) [79] of the lateral chain of poly(4-vinylpyridine) changes the isotherm of the polymer at the air-water interface. Figure 3.13 shows this particular behavior.

The surface behavior of poly(4-vinylpyridine) quaternized with tetradecyl bromide (P4VPC$_{14}$) as function of the quaternization degree has been reported [81]. The percentage of vinylpyridine moieties quaternized was found to be 35–75%. Surface pressure-area isotherms (π – A) at the air-water interface were determined. The polymer monolayer have shown particular shapes at different quaternization degrees. Figure 3.14 shows the (π – A) isotherms of P4VPC$_{14}$.

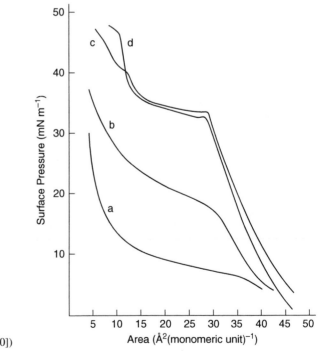

Fig. 3.13 Surface pressure-area isotherms (π – A) of poly(4-vinylpyridine)s quaternized on subphase water at pH 5.7; curve a, P4VPC$_5$Br; curve b, P4VPC$_6$Br, curve c, P4VPC$_8$Br, curve d, P4VPC$_{10}$Br. (From ref. [80])

Fig. 3.14 Surface
pressure-area isotherms
($\pi - A$) of
poly(4-vinylpyridine)
quaternized by tetradecyl
bromide as functions of the
percentage of quaternization.
Subphase water at pH 5.7.
(From ref. [81])

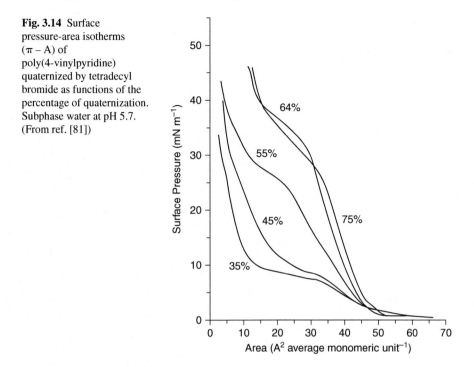

In order to describe what happens in the molecular organization at smaller areas and understand the process undergone by water molecules and counterions under these conditions, it was also developed molecular modeling of a P4VPC$_{14}$ monolayer with full quaternization. Two different surface areas were taken into account. One would correspond to the limiting area A$_0$ (34 Å^2) and another to a larger area (40 Å^2) than the first one. When the compression is beginning to reach 40 Å^2, the lateral chains could be considered almost vertical at the air-water interface. (See Fig. 3.15). The situation shown in Fig. 3.15 seems to favor hydrophobic interaction between them.

According to the radial distribution function (RDF) calculation, the distance between a pyridine group and a Br- counterion should be about 4.7 Å (See Fig. 3.16). When the compression progresses to reach the limiting area, different effects can be seen: First, the side chains present a fair inclination relative to the pyridine group and graphite surface (Fig. 3.15a). Second, the counterions should be close to the pyridine group with respect to the more closed system (3.8 A) (Fig. 3.16). At the same time the solvated counterions wet the aliphatic chains in a significant way (Fig. 3.15a) because of the electrostatic potential produced when the pyridine groups are closed.

In conclusion, in this study [81] it has characterized the Langmuir films formed from P4VPC$_{14}$ with different quaternization degrees spreads on a pure water surface.

This type of polymers called polyelectrolytes are linear macromolecule chains containing a large number of charged or chargeable groups which, in a polar solvent like water, dissociate into charges associated to the polymer backbone and

Fig. 3.15 Molecular model of one monolayer of P4VPC$_{14}$ at air-water interface under periodic boundary conditions: (**a**) 34 Å^2, almost collapse pressure, and (**b**) 40 Å^2. (From ref. [81])

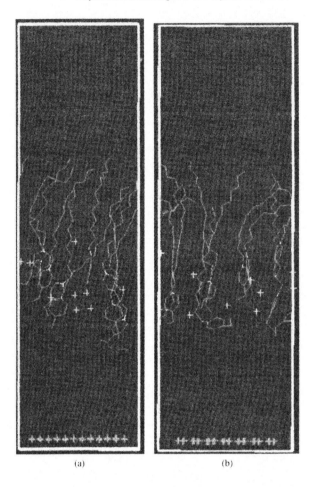

(a)   (b)

counter – ions dispersed in the solution. Charged macromolecules can be used to construct ordered ultrathin solid films by the successive layers of alternating anionic and cationic polymer coating – layer – by – layer deposition (soluble polyelectrolytes) – or by Langmuir – Blodgett technique (insoluble polyelectrolytes).

Most of the studies of insoluble polymer Langmuir monolayers have been devoted to neutral polymers, and less effort have been devoted to polymer monolayers of polyelectrolytes.

Kawaguchi's group [82,83] have pointed out that polymer monolayers of poly(n-alkyl 4- vinylpiridinium) chains show a gas – liquid phase transition at surface pressures below $1 \, N.m^{-1}$. As already mentioned, this behavior is different to that of polymer monolayers of neutral and insoluble polymers.

More recently, Miranda et al. [81] have shown that this family of polyelectrolytes also present a second surface phase transition at surface pressures that depend on the quaternization degree and on the length of the alkyl side chain. Similar results have been presented by Davis et al. [84].

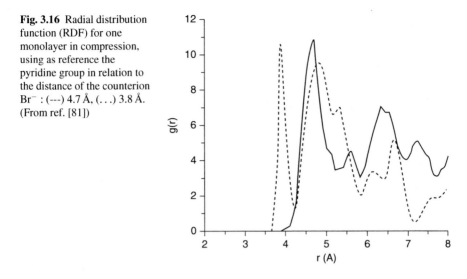

Fig. 3.16 Radial distribution function (RDF) for one monolayer in compression, using as reference the pyridine group in relation to the distance of the counterion Br$^-$: (---) 4.7 Å, (. . .) 3.8 Å. (From ref. [81])

Phase transitions in surface pressure-area ($\pi - A$) isotherms of Langmuir monolayers of various film materials have been the subject of numerous investigations [85, 86].

A characteristic plateau in which there is no or little change in surface pressure upon monolayer compression has been observed with Langmuir monolayers of model amphiphilic compounds such as long chain fatty acids, [85, 87] alkylammonium salts [88] or phospholipids [89]. Figure 3.17 is a good illustration of this behavior. Such a plateau region, which usually disappears at higher subphase temperatures, is attributed to the coexistence of either vapor/liquid phases at low surface pressures, gas phase G in Fig. 3.17, or liquid expanded/liquid condensed phases at high – pressure regions phases C and E in Fig. 3.17.

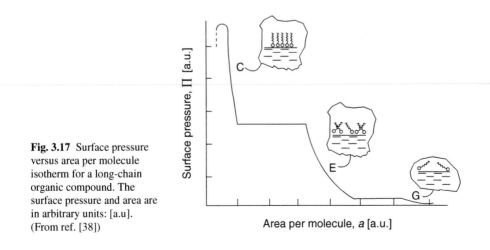

Fig. 3.17 Surface pressure versus area per molecule isotherm for a long-chain organic compound. The surface pressure and area are in arbitrary units: [a.u.]. (From ref. [38])

The relationship between polymer structure and activity at the air – water interface has been also investigated as a function of temperature for various hydrophobically – modified poly (N – isopropylacrylamides) (HM – PNIPAM) [90].

In order to examine the importance of several structural parameters, an unmodified PNIPAM was used as a reference polymer. It is known that poly (N-isopropylacrylamide) (PNIPAM) is a temperature – sensitive water – soluble polymer [91]. Heating an aqueous solution of PNIPAM to a critical temperature ($\sim 32°$) results in a sharp phase separation, at which point the polymer conformation changes from a hydrated coil to a globule with subsequent self – aggregation [92]. It has been observed that on water surface, PNIPAM forms Langmuir monolayers and undergoes also a similar reversible and reproducible phase transition [93]. There is a very interesting behavior the thickness and density of PNIPAM monolayers increase as the subphase temperature approaches the critical phase transition temperature. Modifying PNIPAM with hydrophobic groups the hydrophobic – hydrophilic balance of the PNIPAM backbone was perturbed, and thus it was possible to modulate the phase transition temperature as F. M. Winnik et al. have reported [94, 95].

The interfacial properties of HM – PNIPAM, including the formation and the compression – expansion reversibility of the monolayers, at different subphase temperatures were more recently studied by using the Langmuir film balance technique [90]. The stability and dynamic nature of the HM – PNIPAM monolayers were also further studied by the time – dependent surface pressure measurements. All results have suggested a compression – promoted temperature – and rate – dependent conformational rearrangement of the polymer on the water surface. Increasing the level of hydrophobic modifications progressively improved the monolayer compressibility and stability, and reduced the hysteresis.

The dilational rheology behavior of polymer monolayers is a very interesting aspect. If a polymer film is viewed as a macroscopy continuum medium, several types of motion are possible [96]. As it has been explained by Monroy et al. [59], it is possible to distinguish two main types: capillary (or out of plane) and dilational (or in plane) [59, 60, 97]. The first one is a shear deformation, while for the second one there are both a compression - dilatation motion and a shear motion. Since dissipative effects do exist within the film, each of the motions consists of elastic and viscous components. The elastic constant for the capillary motion is the surface tension γ, while for the second it is the dilatation elasticity ε. The latter modulus depends upon the stress applied to the monolayer. For a uniaxial stress (as it is the case for capillary waves or for compression in a single barrier Langmuir trough) the dilatational modulus is the sum of the compression and shear moduli [98]

$$\tilde{\varepsilon}_K = K + i \omega \eta_k, \tag{3.5}$$

$$\tilde{\varepsilon}(\omega) = \varepsilon_k + \varepsilon_s = \varepsilon_R + i \omega K, \tag{3.6}$$

Figure 3.18 shows different surface relaxation modes and the corresponding viscoelastic parameters [59].

Fig. 3.18 Sketch of different surface relaxation modes and the corresponding viscoelastic parameters. (From ref. [59])

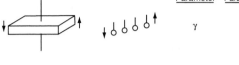

$$\widetilde{\varepsilon}_K = K + i\omega\eta_k, \tag{1}$$

$$\widetilde{\varepsilon}(\omega) = \widetilde{\varepsilon}_k + \widetilde{\varepsilon}_s = \varepsilon_R + i\omega\kappa,$$

	Elastic Parameter	Viscous Parameter
	γ	μ
	ε	κ
	ε_S	κ_S

$$\acute{\varepsilon}_S = S + i\omega\eta_s, \tag{3.7}$$

where ω denotes the angular frequency and the $\omega\eta$'s are the loss components of the compression and shear motions. It is common to refer to ε and κ as the dilatational elasticity and viscosity, respectively [59].

Shear viscosity has been the most intensively investigated property [99].

There are several experimental techniques suitable for studying ε. Some of them are: Relaxation after a sudden compression of the monolayer; Electrocapillary waves; An oscillatory barrier; Light Scattering by thermally excited capillary waves. The first two techniques are used in the low – frequency range, below 1 Hz. The last one in the kilohertz range.

The dilatational rheology of the poly(vinylacetate) monolayer onto an aqueous subphase has been studied between 1°C and 25°C by Monroy et al. [59]. These authors have used the combination of several techniques. By this way, the exploration of a broad frequency range was possible. The relaxation experiments have shown multiexponential decay curves, whose complexity increases with decreasing the temperature. A regularization technique has been used to obtain the relaxation spectra from the relaxation curves and the dilatational viscoelastic parameters have been calculated from the spectra. The shapes of the relaxation spectra agree with the predictions of the theoretical model proposed by Noskov [100].

However, for the temperatures above 15°C the agreement is not quantitative. It is also important to remember that the kilohertz region has been explored by the surface Light Scattering (SLS) technique. The results obtained are compatible with the existence of a single Maxwell mode, with a relaxation time that has an Arrhenius – type temperature dependence. In the intermediate – frequency regime (10 Hz to 2 kHz) a further Maxwell process was found. This behavior might be related to the adsorption – desorption dynamics of loops and tails out of the interfacial plane [59].

It is very well known that the nature of the monolayer partially depends on the strength of interfacial interactions with substrate molecules and that of polymer intersegmental interactions. And it is normal to expect that the viscoelastic properties of polymer monolayer are also dependent on these factors. The static and dynamic properties of several different polymer monolayers at the air – water interface have been examined with the surface quasi-elastic Light Scattering technique combined with the static Wilhelmy plate method [101].

The polymers studied in order of increasing hydrophobicity were poly(ethylene oxide) (PEO), poly(tetrahydrofuran) (PTHF), poly(vinyl acetate) (PVAc), poly (methyl acrilate) (PMA), poly(methyl methacrylate) (PMMA), and poly(tert-butyl methacrylate) (PtBMA). Polymers of varying hydrophobicity are well-known to form expanded and condensed – type monolayers. In this case Kawaguchi et al. [101] have analysed the spectral data in terms of the dispersion equation for capillary wave motion, and the dynamic longitudinal elasticity $\acute{\epsilon}$ and the corresponding viscosity κ of the monolayers as a function of surface concentration were deduced. The static and dynamic elasticities were found to be the same over a majority of the concentration range. For the expanded – type monolayers, $\acute{\epsilon}$ predominates over κ in determining the overall dynamic modulus $\acute{\epsilon}^*$. For the condensed monolayers, PMMA and PtBMA, $\acute{\epsilon}$ and κ contribute about equally to $\acute{\epsilon}^*$. The absolute magnitudes of $\acute{\epsilon}$ and κ increased with increasing hydrophobicity, with PEO the least and PtBMA the greatest [101].

Figure 3.19 shows the variation of the frequency shifts f_S, a measure of the propagation velocity of the capillary waves, as a function of surface concentration Γ, which is the reciprocal A, for different polymers at a wavenumber k = 323 cm^1.

Another interesting result found by Kawaguchi et al. [101] was that the elasticities of PEO and PTHF are quite similar, however, κ for PTHF is two to three times larger. PEO is very expanded on the surface and the low κ will be probably due to a very low degree of cohesion. PVAc and PMA were almost identical relative to the magnitudes of ϵ and κ, which were slightly larger than those for PTHF. Since PMMA and PtBMA have presented the most significant viscous contribution to the dynamic modulus, and this can be due to the large segment – segment cohesion in the monolayer state while PEO will have the smallest segment – segment cohesion. Another interesting comparison deals with the collapse of the expanded monolayers which was monitored via ϵ and κ at high concentrations. PEO and PTHF are well above their bulk glass transition point at 25°C, while PVAc and PMA are close to their glass transition point at this temperature. It may be that this is why it was possible see the large increase in κ after monolayer collapse for PVAc and PMA, while for the more liquid like polyether monolayers κ does not change as the monolayer collapses [101].

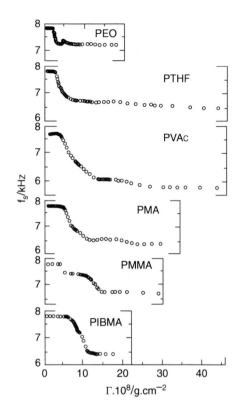

Fig. 3.19 Plots of frequency shift f versus surface concentration Γ for six polymer monolayers at a scattering wavevector $k = 323\,\text{cm}^{-1}$. (From ref. [101])

3.3 Amphiphilic Block Copolymer Behavior in Solution and Interfaces

Amphiphilic block copolymers form an important class of polymeric materials which have attracted considerable attention because of their outstanding solution properties and a wide range of applications [102, 103]. These materials are very interesting from the point of view of fundamental research, as they exhibit self – assembling properties in the presence of a selective solvent or surface [104–106]. Surface micelles at the air – water interface, of different morphologies depending on the balance between block sizes, have been identified by transmission electron microscopy (TEM) and atomic force microscopy (AFM) for block polyelectrolytes [107] and for nonionic diblock copolymers [108]. The term "surface micelle" is used in the sense described by Langmuir [109]. However, this term is often used in a wider sense and seems to refer to a number of different phenomena. In a more general term "surface aggregate" it will be use here. Surface aggregation has been found to be general for a number of block copolymers, including quaternized poly(4-vinylpyridine) (P4VP), poly(tert-butyl acrylate) (PtBA) and poly(dimethylsiloxane) (PDMS) [110, 111].

Fig. 3.20 Nanoparticulate drug delivery systems formed by amphiphilic block copolymers and their general characteristics. (From ref. [112])

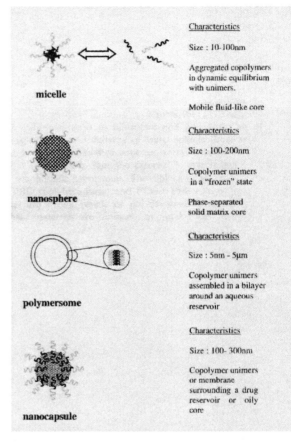

Block Copolymers are macromolecules which are composed of blocks usually in linear as it shown in Fig. 3.20, where it is illustrated a classical block copolymer. Main block copolymers are amphiphilic block copolymers having united hydrophilic blocks to hydrophobic blocks. Amphiphilic block copolymer have surfactant properties and form different kinds of associations, such as micelles, nanospheres, nanocapsules and polymersomes This tipe of association can act like excellent vehicles of several active principles. The composition, aggregate formation and the different applications of these materials have been reviewed [112].

Figure 3.20 also illustrates the nanoparticulate drug delivery systems formed by amphiphilic block copolymers and their general characteristics.

One of the most fascinating properties of block copolymers is their ability to self-assemble into ordered nanostructures not only in selective solvent and melts but also at the interface and surface [113]. Self-assembly of amphiphilic or surface-adsorbing block copolymers at the air-water interface can form two dimensional monolayers on the nanometer scale order.

Amphiphilic block copolymers have attracted a great deal of attention in terms of their ability to form nanoparticles. As normally one type of segment, is hydrophobic and the other is hydrophilic, so the resulting macromolecule is composed of regions which have opposite affinities for an aqueous solvent. These materials, when intended for use in drug delivery, are generally composed of biocompatible, biodegradable hydrophobic polymer blocks such as polyesters or poly(amino acids) covalently bonded to a biocompatible hydrophilic block. The literature abounds with studies using amphiphilic block copolymers of different compositions and various methods of preparation that produce nanoparticles referred to as micelles, nanospheres, core-shell nanoparticles, micelle-like nanoparticles, crew cut micelles, nanocapsules and polymersomes.

There is aggregate formation if an amphiphilic block copolymer dissolves in a liquid that acts as a good solvent for one of the blocks and a bad solvent for another. Their macromolecules can be associate to form aggregates similar to those surfactants obtained with low molecular mass.

Due to the unique structure of amphiphilic macromolecules they have a tendency to accumulate at the boundary of two phases and thus they present a surfactant properties. In aqueous solutions, amphiphilic macromolecules orientate themselves so that the hydrophobic blocks are removed from the aqueous environment in order to achieve a state of minimum free energy. As the concentration of macroamphiphile in solution is increased, the free energy of the system begins to rise due to unfavourable interactions between water molecules and the hydrophobic region of the macroamphiphile resulting in structuring of the surrounding water and a subsequent decrease in entropy. At a specific and narrow concentration range of macroamphiphile in solution, termed the critical micelle concentration (CMC), several macroamphiphiles will self-assemble into colloidal-sized particles termed micelles. The formation of micelles effectively removes the hydrophobic portion of the macroamphiphile from solution minimizing unfavourable interactions between the surrounding water molecules and the hydrophobic groups of the amphiphile. Micelles typically have diameters ranging from 10 to 100 nm and are characterized by a core-shell architecture in which the inner core is composed of the hydrophobic regions of the amphiphiles creating a cargo space for the solubilization of lipophilic drugs. The core region is surrounded by a palisade or corona composed of the hydrophilic blocks of the amphiphiles. The hydrophilic blocks forming the corona region become highly water bound and adopt a "splayed" appearance, giving rise to different conformations such as a polymer brush [113].

A polymeric nanosphere may be defined as a matrixtype, solid colloidal particle in which drugs are dissolved, entrapped, encapsulated, chemically bound or adsorbed to the constituent polymer matrix .These particles are typically larger than micelles having diameters between 100 and 200 nm and may also display considerably more polydispersity (Fig. 3.21).

If the core of the vesicle is an aqueous phase and the surrounding coating is a polymer bilayer, the particle is referred to as a polymersome (Fig. 3.22).

Polymeric vesicles, or polymersomes, are of interest for the encapsulation and delivery of active ingredients. They offer enhanced stability and lower permeability compared to lipid vesicles, and the versatility of synthetic polymer chemistry provides the ability to tune properties such as membrane thickness, surface

Fig. 3.21 Copolymer
unimers in a "frozen" state.
(From ref. [112])

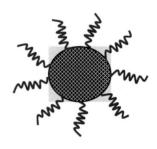

functionality, and degradation kinetics. One approach to form large polymersomes with diameters of 10–100 nm is to use water-in-oil-in-water double emulsion drops of controlled architecture as templates. A volatile organic solvent containing an amphiphilic diblock copolymer is used as the middle phase; evaporation of the solvent leads to polymersome formation, as shown schematically in Fig. 3.23. This technique offers the advantages of high encapsulation efficiencies and controllable vesicle sizes and architectures.

The formation of polymersomes from water in- oil-in-water drops. Initially, a double emulsion consisting of single aqueous drops within drops of a volatile organic solvent ("oil") is prepared using a microcapillary device. Amphiphilic diblock copolymers dissolved in the middle phase assemble into monolayers at the oil-water interfaces. Evaporation of the solvent then leads to the formation of polymer bilayers (polymersomes).

When the core is an oily liquid, the surrounding polymer is a single layer of polymer, and the vesicle is referred to as a nanocapsule. These systems have found utility in the encapsulation and delivery of hydrophobic drugs Polymers used for the formation of nanocapsules have typically included polyester homopolymers such as poly(D,L-lactic acid) (PLA), poly(lactic-co-glycolic acid) (PLGA) and poly(caprolactone) PCL [112].

Amphiphilic polymers have been applied in numerous surface technologies and reviewed in various aspects. These polymers are polyampholytes, which contain, as it was remarked, both hydrophilic and hydrophobic components in their macromolecules. Their peculiar adsorption behavior is another interesting and fundamental issue [112].

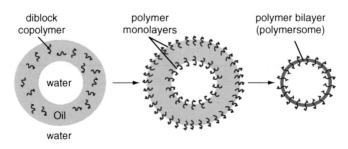

Fig. 3.22 Schematic illustration of the formation of polymersomes. (From ref. [148])

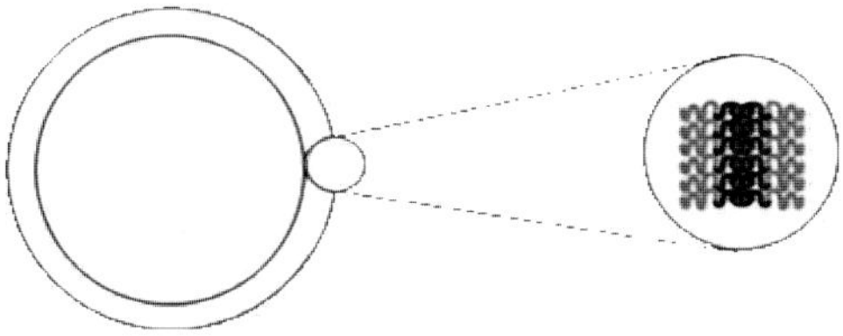

Fig. 3.23 Copolymer
unimers assembled in a
bilayer around an aqueous
reservoir. (From ref. [112])

There has been considerable recent interest in the self − assembly and surface activity of amphiphilic polymers and copolymers. Their interfacial and bulk solution properties have shown a rich pattern of behavior, and the ability to tailor their properties offers a wide range of potential applications. Their bulk aggregation behavior make them candidates, for example, for dye transportation and drug delivery; whereas their surface properties make them useful as colloid stabilisers, anti − foaming agents and emulsifiers. This behavior can be illustrated in Fig. 3.24.

Colloid stabilization with amphiphilic polymers [2, 114, 115] requires the formation of a thick polymer layer around each particle in order to create a repulsive steric force that overcomes the van der Waals attraction. This is usually done by adsorbing on the colloidal particle a polymer solution in a good solvent, which builds up on the surface a fluffy layer with a thickness of the order of the radius of gyration of isolated polymer chains, in general of the order of a few hundred angstroms.

Marques et al. [116] have studied the adsorption of an A − B diblock copolymer from a dilute solution onto a solid surface that attracts the A block and repels the B block in a nonselective solvent, good for both blocks.

Another of the most fascinating properties of block copolymers is their ability to self-assemble into micelles, aggregates, and vesicles of various morphologies in the presence of a selective solvent, [117, 118] and recent studies have demonstrated that self- assemble of amphiphilic block copolymers into various morphologies occurs not only in selective solvents but also at interfaces and surfaces [119, 120].

In the case of amphiphilic or surface − adsorbing block copolymers, the self − assemble structure at the air-water interface can also transferred to a solid substrate using the Langmuir − Blodgett (LB) transfer technique [121–123].

A schematic diagram illustrating the commonest form of LB film deposition is shown in Fig. 3.25. In this example the substrate is hydrophilic and the first

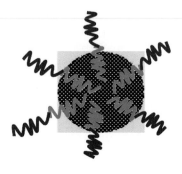

Fig. 3.24 Copolymer
unimers or membrane
surrounding a drug reservoir
or oily core. (From ref. [112])

Fig. 3.25 Langmuir –
Blodgett Transfer Technique.
Y-type Langmuir- Blodgett
film deposition. (From
ref. [23])

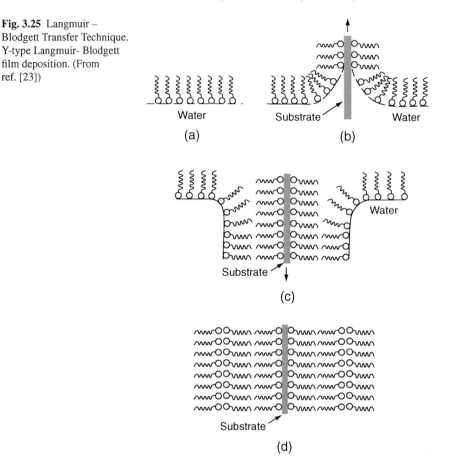

monolayer is transferred, like a carpet as the substrate is raised through the sur-
phase (Fig. 3.25b), the substrate may therefore be placed in the subphase before the
monolayer is spread or may be lowered into the subphase through the compressed
monolayer. Subsequently a monolayer is deposited on each traversal of the surface
(Fig. 3.25c). as shown in Fig. 3.25d, these stack in a head-to-head and tail-to-tail
configuration, this deposition mode is referred to as "Y" type.

The interfacial properties of an amphiphilic block copolymer have also attracted
much attention for potential functions as polymer compatibilizers, adhesives, colloid
stabilizers, and so on. However, only a few studies have dealt with the monolayers
o well – defined amphiphilic block copolymers formed at the air – water interface.
Ikada et al. [124] have studied monolayers of poly(vinyl alcohol)- polystyrene graft
and block copolymers at the air – water interface. Bringuier et al. [125] have studied
a block copolymer of poly (methyl methacrylate) and poly (vinyl-4-pyridinium bromide) in order to demonstrate the charge effect on the surface monolayer- forming
properties. Niwa et al. [126] and Yoshikawa et al. [127] have reported that the poly
(styrene-co-oxyethylene) diblock copolymer forms a monolayer at the air – water

interface, where the poly (styrene) hydrophobic block was monomolecularly aggre-
gated on the water surface [128]. In this case, the water – soluble poly(oxyethylene)
block of the copolymer can be considered as a "tethered" polymer chain [102],
anchored to the interface, and the segmental concentration profile of which was
investigated by surface pressure and neutron reflectivity measurements.

An amphiphilic diblock copolymer spread from a solution of organic solvent
onto the water surface, normally were found to form a stable monolayer [129].
The surface monolayer has been successfully transferred onto a substrate by the
Langmuir – Blodgett technique. Some times the surface pressure – area isotherms
exhibited a plateau region, suggesting a structural change taking place on the water
surface at specific pressures.

The observed values of the layer thickness and the occupied area can indicate
that the hydrophobic segments, which lie essentially flat on the water surface at
low surface pressures, aggregate at higher pressures, forming a thicker layer with
hydrophobic and hydrophilic segments microphase – separated at the air – water
interface. This behavior can be illustrated in Figs. 3.26 [129] and 3.27 [130] for
different diblock copolymers.

Amphiphilic block copolymers consisting of a hydrophobic (poly(ethyl ethylene)
(PEE) and a hydrophilic poly(ethylene oxide)(PEO) block form monolayers at the
air-water interface. The schematic molecular arrangement of this diblock is shown
in Fig. 3.26.

Figure 3.27 shows the π – A isotherms, where the surface pressure π increases
in two steps with decreasing A, the occupied surface area per chain, giving a plateau
region [130].

Table 3.7 summarizes the molecular characteristics of the Block copolymers
shown in Fig. 3.28. The homopolymer is poly(isobutyl vinyl ether) (PIBVE). P1 and
P2 are the block copolymers of isobutyl vinyl ether (IBVE) and a vinyl ether with
a protected glucose residue (3-O-(vinyloxy) ethyl-1,2:5,6-di-O-isopropylidene-D-
glucofuranose).

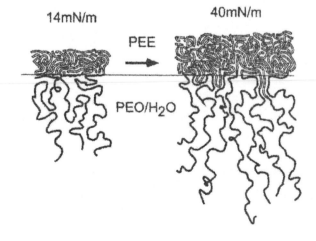

Fig. 3.26 Schematic of the
molecular arrangement of
PEE-PEO monolayer and its
changes on compression.
(From ref. [129])

Fig. 3.27 Schematic illustration of monolayers at the air-water interface. (From ref. [130])

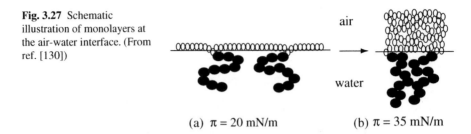

(a) π = 20 mN/m   (b) π = 35 mN/m

Table 3.7 Molecular characteristics of the Block copolymers and homopolymer. (From ref. [130])

Sample code	m/n	$M_n{}^a$	M_w/M_n
P1	20/48	9800	1.04
P2	20/89	13900	1.06
PIBVE	0/39	3900	1.07

^{a} Calculated from the block composition (m/n).

The dramatic morphological changes were observed in the Langmuir – Blodget (LB) film assemblies of poly (ethylene glycol) – b –(styrene-r-benzocyclobutene) (PEG-b-(S-r-BCB)) after intramolecular cross – linking of the S-r-BCB block to form a linear – nanoparticle structure by Kim et al. [131]. In this investigation one of the problem was to clarify if the block copolymer exits as individual molecules or as surface aggregates at the air – water interface. To answer this question, they have considered the surface viscoelastic properties of the block copolymers measured using the interfacial stress rheometer.

The result was interpreted under two scenarios in which 1) discrete molecules or 2) surface aggregates of the block copolymers should exist at the air – water interface at different compression states.

Combining the surface dynamic moduli measurements with the morphologies of the LB transferred block copolymer films imaged by atomic force microscopy

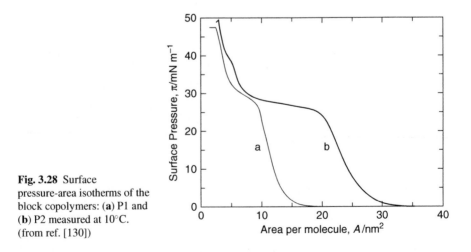

Fig. 3.28 Surface pressure-area isotherms of the block copolymers: (**a**) P1 and (**b**) P2 measured at 10°C. (from ref. [130])

(AFM) as a function of different compression state, they have used a modified approach to interpreted the $\pi - A$ isotherms of these amphiphilic block copolymers in which they took into account the presence of the surface aggregates at the air – water interface.

These authors have considered such interpretation more appropriate for the block copolymers under study as they have observed markedly different surface rheological behaviors between the linear and the linear-nanoparticle block copolymers, which could not be explained if the block copolymers existed as discrete molecules at the air – water interface. Although the focus of this work was on the morphological change of the surface aggregates promoted by the architectural difference of the macromolecules, it was essential to first establish that the surface aggregation process was occurring at the air – water interface rather than on the solid substrate after the LB transfer to understand how the different morphologies arise from the different block copolymer architectures [131].

Physicochemical properties of (poly(γ-benzyl-L-glutamate)) (PBLG) –(poly (ethylene oxide)) (PEO) diblock copolymers composed of the hydrophobic rod component and PEO as the hydrophilic component have been investigated at the air – water interface [132]. Rod – coil molecules consisting of a rigid rod block and a flexible coil block are a novel type of block copolymer with a unique microstructural organization. Hydrophobic and hydrophylic effects, electrostatic interactions, and hydrogen – bonding control the phase behavior of these rod – coil molecules. It is known that polymers with a stiff helical rodlike structure have many advantages over other synthetic polymers because they often possess stable secondary structures due to cooperative intermolecular interactions. Then, the incorporation of an elongated coil like block into the helical rod system in a single molecule is a unique way to create new supramolecular structures. Park et al. [132] taking in account that the monolayer at the air – water interface provide an important and convenient model system for investigation the behavior of rod – coil copolymers, have studied the monolayer behavior of PBLG – PEO diblock copolymers at the air – water interface. The block copolymers had the same hydrophilic PEO chains [133] and different PBLG chain lengths.

The surface pressure – area isotherms were collected at different temperatures and the energy relationship between rod and coil as a function of rod length was analized [134]. The microstructures of these monolayers based on the energy relationship were also investigated using Atomic Force Microscopy (AFM).

The results show that the energy relationship of PBLG – PEO diblock copolymers is a competition between entropy loss of the PEO aggregation and enthalpy decrease of the PBLG packing. When the enthalpy decrease of the PBLG packing is smaller than the entropy loss, the surface pressure presents a positive temperature coefficient. When the enthalpy decrease of the PBLG packing is greater than the entropy loss, the observed surface pressure has a negative temperature coefficient.

The enthalpy effect of rod packing influences the microstructures at the air- water interface. The copolymer with the long rods, forms a cylindrical structure in the monolayer due to the bigger enthalpy decrease. Copolymer, with the middle-sized rods, will form micellar structure in the monolayer by self – assembly. Copolymer with the short rods, will be a bilayer structure in the monolayer and clusters with nodal self – assembly. These results according to Park et al. [132] have demonstrated that the microstructures of PBLG – PEO diblock copolymers are related to energy differences between rods and the coil block.

Figure 3.29 shows atomic force microscopy (AFM) images for the three copolymers, GE-1, GE-2, and GE-3 at low and high concentrations.(From ref. [132]).

The characteristics of the block polymer PBLG – PEO were summarized in Table 3.8.

Amphiphilic block copolymers can be also considered as a particular class of surfactant [135]. They are composed, as it is very well known, of at least one hydrophobic block and one hydrophilic block, and for this reason, they are usually called polymeric surfactants or "macrosurfactants". In comparison to classical surfactants,

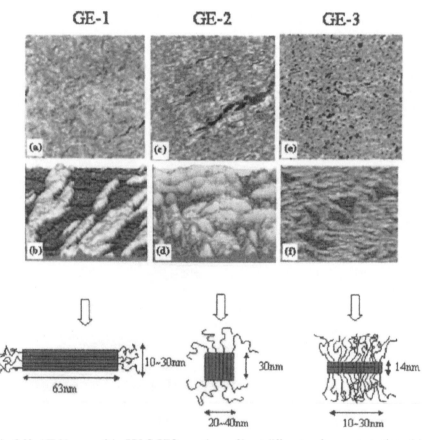

Fig. 3.29 AFM images of the PBLG-PEO monolayers film at different surface concentrations: (**a**) GE-1 at high concentration (3.0 mg/m², 5 × 5 μm), (**b**) GE-1 at low concentration (1.5 mg/m², 250 × 250 nm); (**c**) GE-2 at high concentration (3.0 mg/m², 5 × 5 μm), (**d**) GE-2 at high concentration (1.5 mg/m², 200 × 200 nm) (**e**) GE-3 at high concentration (2.5 mg/m², 5 × 5 μm), (**f**) GE-3 at low concentration (1.5 mg/m², 250 × 250 nm). (From ref. [132])

Table 3.8 The molecular weight and the monomer ratio of GE-1, GE-2 and GE-3. (From ref. [132])

Polymer	Extrapolated area[a] (nm^2)	Supposed PBLG area[b] (nm^2)
GE-1	77.65	82
GE-2	67.18	35
GE-3	35.74	7.4

[a] Extrapolated area was obtained form the $\pi - A$ isotherm.
[b] BLG area is 0.197 nm^2/residue and the PEO area varied with compression at the air–water interface.

amphiphilic block copolymers often exhibit a reduced mobility and slower diffusion rates [136]. As a direct consequence, the equilibrium between polymeric micelles can take many days [137]. Moreover, macrosurfactants have much lower critical micelle concentrations (CMC) than their low – molecular- mass counterparts [138].

Normally, the CMCs of macrosurfactants have been reported in the concentration range from 10^{-9} to 10^{-4} mol $\times$ L^{-1} [136, 139] whereas common surfactants such as sodium dodecyl sulfate (SDS) or another as cetyltrimethylammonium bromide (CTAB) exhibit CMCs on the order of 10^{-3} to 1 mol $\times$ L^{-1} [140]. The CMC might be absent for macrosurfactants [141].

The extremely low CMCs have been advantageous for several applications, since only traces of polymer are required to form micelles. High dilution effects, that are problematic in the case of classical surfactants, do not alter polymeric micelles. The surface activity at the air – water, of the amphiphilic block copolymer or polymeric surfactants must be different from the classical surfactants, because of their much lower diffusion coefficients and their much complex conformations.

3.4 Polymer Adsorption from Solution

Polymer adsorption from solution is a very large subject and it is difficult to provide an exhaustive treatment. We will try to describe the scaling and self- consistent field descriptions of homopolymer adsorption, together with experimental data selected to illustrate the important aspects.

The free surface energy of a polymer solution may attract the polymer which forms an adsorbed layer at the surface. The number of monomers per unit volume is a function C (x) of the distance from the surface. Figure 3.30 illustrates this situation [142].

This function was studied assuming that the polymers are long in a good solvent; the surface is strongly attractive, and the solution itself is so dilute that $C (x \rightarrow \infty) = 0$. Several times, P.G. de Gennes [143] has considered this situation and shown [97] that C(x) decreases like $x^{-4/3}$ (in dimension three). Being x is a distance from the surface. The argument given by him can be explained as follows. The monomer concentration for long polymers in a good solvent (kuhnian chains) is proportional to the number C of polymers per unit volume and in the kuhnian limit, it has to be represented by the following expression

Fig. 3.30 An adsorbed
polymer near a surface. The
length "a" defines the range
in which the surface potential
can be felt. Beyond that
range, there are only
repulsive forces between
polymers and this repulsion
creates a self-consistent
potential. (From ref. [142])

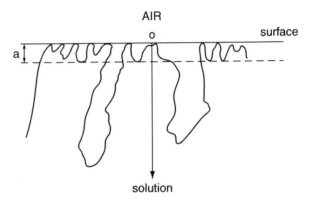

$$C_k = CX^{1/\nu} \tag{3.8}$$

where $R^2 = X^2 d$ is the mean square distance between the end points of a isolated polymer and ν the well known size exponent or critical size exponent ($\nu = 0.59$). The concentration C_k can be easily measured; type of concentration can be considered as important from any point of view.

Near a free surface, the quantity of interest for kuhnian chains is the concentration $C_k(x)$ which obviously has the same dimension as C_k namely $C_k \sim L^{-d+1/\nu}$. Therefore as no other length occurs in the problem, it is possible to have

$$C_k(x) = A x^{-d+1/\nu} \tag{3.9}$$

Where A is a pure number.

It must emphasize the fact that, in the problem under consideration, $C_k(x)$ vanishes when $x \to \infty$, and in the absence of any cut – off, would become infinite on the surface; moreover the chains are assumed to be very long. This is the reason why it is impossible to introduce any physically meaningful correlation length in the problem; the consequence of this fact is that equation (3.9) is only universal law that can be written in the present situation.

J. des Cloizeaux claims that A is a universal number which depends only on the dimension of space (A = A (d). Thus C (x) is not only given by a self – similar expression as was pointed out by P.G. de Gennes [143], it is also completely determined.

It may look surprising that A should not depend on the strength of the attraction but the explanation can be understood if we considered that the polymers that are adsorbed on the surface repel one another. Adsorption from the solution takes place until the repulsion compensates the attraction of the surface. By this way, in all the cases, a polymer that is adsorbed on the surface depend is only marginally bound. In spite of the fact that A should be universal, the quantity of polymer that is adsorbed on the surface depend on the properties of the system. This corresponds to the fact that the integral

$$\int_{0}^{\infty} dx \, x^{-d+1/v} \qquad (3.10)$$

is divergent at $x = 0$. Then, in the vicinity of the surface, the concentration $C_k(x)$ is represented by a function of the form

$$C_k(x) = x^{-d+1/v} f(x/a) \qquad (3.11)$$

Where a is a cut-off corresponding to the range of the attractive forces. It is possible, for instance assume that $f(x/a) = A$ for $x \geq a$. In this case, the number of monomers per unit surface is given by

$$\sigma = \int_{0}^{\infty} dx \, C_k(x) = a^{-(d-1-1/v)} \left[\int_{0}^{1} dt \, f(t)/t^{d-1/v} + A/d - 1 - 1/v \right] \qquad (3.12)$$

(for $v > 1/d - 1$)

By this way, it can understand why A is universal and at the same time, for $d = 1$ the result is trivial. In this case, a long straight polymer starts from the surface ($v = 1$) and it has $C_k(x) = 1$, $A(1) = 1$.

It is also shown that a mean field theory is valid for small values of $\varepsilon = 4 - d$ and that it leads to the result

$$A(4 - \varepsilon) = 2/\pi^2 \, \varepsilon, \quad \text{for } 0 < \varepsilon \ll 1. \qquad (3.13)$$

For small values of $\varepsilon = 4 - d$, a polymer chain in solution is nearly Brownian and a mean field method might reasonable results in this limit. Thus it is possible consider that the chains feel a potential $V(x)$ which is the sum of the (attractive and repulsive) surface potential and of a self – consistent potential produced by the other chains.

The result is interesting and the fact that $A(4 - \varepsilon)$ becomes infinite when $\varepsilon \to 0$ can be easily understood. When ε is very small, the chains are nearly Brownian, and they repel very weakly one another; thus, when $\varepsilon \to 0$, more and more chains are attracted by the surface.

Unfortunately, as des Cloizeaux has noted, for practical applications, the preceding result is not very useful. It is necessary to calculate the next order in ε.

Another interesting aspect is the conformational changes when the chains are under different concentration conditions: isolated dilute or semidilute or melt. The adsorbed layer thickness increases with increasing concentration, mainly due to the contribution of tails. Figure 3.31 shows this behaviour.

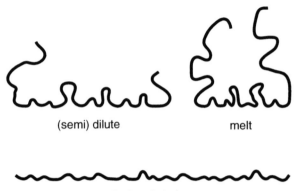

(semi) dilute   melt

isolated chains

Fig. 3.31 Schematic picture of average adsorbed chain conformations in extremely dilute solution (isolated chains on the surface), dilute and semidilute solutions, and the polymer melt. The adsorbed layer thickness increases sharply with increasing concentration, mainly due to the contribution of tails. Significant tail formation occurs as soon as the chains begin to compete for surface sites. (From ref. [144])

3.5 Wettability Behavior and Contact Angles

Wetting of a solid by a liquid is normally described in terms of the equilibrium contact angle θ and the appropriate interfacial tensions as shown in Fig. 3.32.

At equilibrium, equation the forces leads to Young's equation,

$$\gamma_{S/V} = \gamma_{S/L} + \gamma_{L/V} \cos \theta, \tag{3.14}$$

where $\gamma_{S/V}$, $\gamma_{S/L}$ and $\gamma_{L/V}$ are the interfacial tensions at the solid/vapour, solid/liquid and liquid/vapour interfaces respectively. In all the cases the vapour refers to that of the liquid, i.e., the system is at the equilibrium with the vapour at its saturated vapour pressure. It is important and necessary to remember that Young's equation only applies to a system at equilibrium and for which $\gamma_{L/V}$ and θ are given their equilibrium values. For practical purposes, a liquid does not wet a solid when $\theta > 90°$, although strictly speaking a zero contact angle signifies wetting and the complete and spontaneous displacement of air from the surface.

From Young's equation

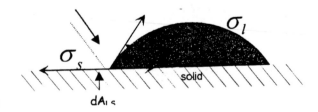

Fig. 3.32 Forces acting on a
rop resting on a solid surface

$$\cos\theta = \gamma_{S/V} - \gamma_{S/L}/\gamma_{L/V} \tag{3.15}$$

from which it is apparent that if $\theta < 90°$, a decrease in $\gamma_{L/V}$ will reduce θ and hence improve the wetting. The addition of a surface active agent causes a reduction in $\gamma_{L/V}$ and, if adsorbed, a change (probably a decrease) in $\gamma_{S/L}$, both effects leading to better wetting. The change in $\gamma_{S/}$ is probably negligible in most cases; the dominating factor in wetting is normally $\gamma_{L/V}$.

The work Wd involved in the wetting of 1 cm^2 of the external surface of a powder by a liquid is given by the difference between the interfacial energies before and after wetting

$$Wd = \gamma_{S/L} - \gamma_{S/V} \tag{3.16}$$

$\gamma_{S/V}$ is the surface tension of the solid in equilibrium with the vapour of the liquid and following Patton [145], Wd is termed the work of dispersion.

The wettability of a polymer film normally is determined by static contact angle measurements. The surface free energy (SE) of a polymer can be determined by wettability measurements with two different liquids. The dispersion force and polar contributions to SE, γ_D and γ_P respectively, are also calculated normally by using the Owens and Wendt, and Kaelble methods [146, 147]. The measurements of contact angles (CA) on a given solid surface is one of the most practical ways to obtain surface free energies.

References

1. E.D. Goddard, and B. Vincent (Eds), Polymer Adsorption and Dispersion Stability; Am. Chem., Sec. Symps, Series 240, 227 (1984)
2. D.H. Napper, "Polymeric Stabilization of Colloidal Dispersion", Academic Press, London (1983)
3. A. Leiva, A. Farias, L. Gargallo, D. Radić, Eur. Polym. J., 44, 2589 (2008)
4. R.Jones, R. Richards, "Polymers at Surfaces and Interfaces", Cambridge University Press (1999)
5. T. Maruyama, J. Langer, G.G. Fuller, C.W. Frank, C.R. Robertson, Langmuir, 14, 1836 (1998)
6. A. Leiva, L. Gargallo, D. Radić, J. Macromol. Sci. Part A: Pure Appl. Chem. 41 (5), 577 (2004)
7. A. Asnacias, D. Langevin, F.J. Argillier, Macromolecules, 29, 23, 7412 (1999)
8. P.F. Luckham, "Polymer Surfaces and Interfaces", W. J-Feast and H.S. Munro, John Wiley and Sons Ltd (1987)
9. L. Gargallo, D. Radić, Curr. Trends Polym. Sci., 6, 121–133 (2001)
10. G.J. Fleer, J.M.H.M. Scheutjens, Adv. Colloid Interface Sci., 16, 341 (1982)
11. A.W. Adamson, A.P. Gast, "Physical Chemistry of Surfaces" 6th Ed. Wiley Interscience, New York 1997
12. F.J. Holly, M.F. Refojo, J. Biomed. Mater. Res. 9, 315 (1975)
13. B.D. Ratner, J. Appl. Polym. Sci. 22, 643 (1978)
14. B.D. Ratner; in "Biomaterials: Interfacial Phenomena and Applications" Cooper and Peppas, Eds., ACS Advances in Chemistry Series Vol. 199. Washington, DC, (1982)

15. L. Lavielle, J. Schultz, J. Colloid Interface Sci. 106, 438 (1985)
16. J.D. Andrade, D.E. Gregonis and L.M. Smith in "Physicochemical Aspects of Polymer Surfaces", K.L. Mittal, Ed. Vol. 2, p. 911 Plenum, New York (1983)
17. E. Ruckenstein, S.V. Gourisankar, J. Colloid Interface Sci. 109, 557 (1985)
18. S.H. Le, E. Ruckenstein, J. Colloid Interface Sci. 117, 172 (1987)
19. D.T. Clark, A. Dilka, D. Shuttleworth, in "Polymer Surfaces", D.T. Clark and W.J. Feast, Eds. P. 185. Wiley, New York (1978)
20. H. Yasuda, H.C. Marsh, S. Brandt and C.N. Reilley, J. Polym. Sci. Polym. Chem. Ed. 15, 991 (1977)
21. A. Baszkin, M. Nishino, L. Ter-Minassian-Saraga, J. Colloid. Interface Sci. 59, 516 (1977)
22. B.D. Ratner, in "Polymer Science and Technology ", Lee, Ed. Vol. 12B p. 691. Plenum, New York (1980)
23. A. Leiva, M. Urzua, L. Gargallo and D. Radić, J. Colloid Interface Sci. 299, 70 (2006)
24. B. Miranda, L. Gargallo, A. Leiva, M. Urzúa, F. González-Nilo, D. Radić, Polymer, 44, 3969 (2003)
25. S.H. Lee, E. Ruckenstein, J. Colloid Interface Sci. 120, 529 (1987)
26. A. Dilks., Anal. Chem., 53, 802A (1981)
27. D. Clark, W. Feast, Polym. Surf., Wiley, New York (1978)
28. S.L., Regen, P. Kirszensztejn, A. Singh, Macromolecules, 16, 335 (1983)
29. R.J. Vig, J. Vac. Sci. Tech., A3, 1027 (1985)
30. R. Foerch, N.S. McIntyre, D.H. Hunter, J. Polym. Sci. Part A. Polym. Chem., 28, 803 (1990)
31. G. Gillberg, J. Adhesion, 21, 129 (1987)
32. F. MacRitchie, J. Colloid Interface Sci., 105, 1, 119 (1985)
33. I. Langmuir, D.F., Waugh, J. Am. Chem. Soc., 62, 2771 (1940)
34. G. González, F. MacRitchie, J. Colloid Interface Sci., 32, 55 (1970)
35. F. MacRitchie and L. Ter-Minassian-Saraga, Colloids Surf., 10, 53 (1984)
36. I. Langmuir, Science, 87, 493 (1938)
37. K.B. Blodgett, J. Am. Chem. Soc., 57, 1007 (1935)
38. M.C. Petty, "Langmuir-Blodgett Films", Cambridge University Press, Cambridge (1996)
39. D. Crisp, in "Surface Phenomena in Chemistry and Biology", Pergamon, New York (1958)
40. N. Beredjik, in "Newer Methods of Polymer Characterization", Interscience, New York, (1964)
41. G.L. Gaines, Jr., "Insoluble Monolayers at Liquid – Gas Interfaces, Wiley, New York (1966)
42. G.L. Gaines, Jr., Langmuir, 7, 834 (1991)
43. A. Ulman, "An Introduction to Ultytrathin Organic Films from Langmuir – Blodgett to Self – Assembly", Academic Press, New York (1991)
44. L. Gargallo, A. Leiva, L. Alegria, B. Miranda, D. Radić, J. Macromol. Sci. Phys., B43, 5, 913 (2004)
45. M. Heskin, J.E. Guillet, J. Macromol. Sci. Chem. 2 (8), 1441 (1968)
46. L. Gargallo, B. Miranda, A Leiva, A. Gonzalez, C. Sandoval, D. Radić, L.H. Tagle, J. Macromol. Sci. PartB: Phys., 45, 105 (2006)
47. J.A. Bergeron, G.L. Gaines, Jr., W.D. Bellamy, J. Colloid Interface Sci. 25, 97 (1967)
48. T. Yamashita, Nature, 231, 445 (1971)
49. G. Gabrielli, M. Pugelli, R. Faccioli, J. Colloid Interface Sci. 37, 213 (1971)
50. G. Gabrielli, A. Maddij, J. Colloid Interface Sci., 64, 19 (1978)
51. M. Kawagushi, S. Komatsu, M. Matsuzumi, A. Takahashi, J. Colloid Interface Sci., 102, 356 (1984)
52. S.Y. Mumby, J.D. Swalen, J.F. Rabolt, Macromolecules, 19, 1054 (1986)
53. J. Kumaki, Macromolecules, 19, 2258 (1986)
54. M.A. Noordegraaf, G.J. Kuiper, A.T.M. Marcelis, E.J.R. Sudholter, Macromol. Chem. Phys. 198, 3681 (1997)
55. F. Davis, P. Hodge, C.R. Towns, Z. Ali-Adib, Macromolecules, 24, 5695 (1991)
56. M.C. Petty, "Polymer Surfaces and Interfaces", Ed. by W.J. Feast and H.S. Munro, John Willey and Sons, Chichester (1987)

57. K. Gong, S. Feng, M.L. Go, P.H. Soew, Coll. Surf. A: Physiochem. Eng. Aspects 207, 113 (2002)
58. C. Ringard-Lefebvre, A. Baszkin, Langmuir, 10, 2376 (1994)
59. F. Monroy, F. Ortega, R.G. Rubio, Phys. Rev. E, 58, 6, 7629–7641 (1998)
60. F. Monroy, F. Ortega, R.G.Rubio, Eur. Phys. J., B, 13, 745 (2000)
61. P.G. de Gennes, Scaling concepts in Polymer Physics, Cornell University Press, Ithaca, NY (1979)
62. P.G.de Gennes, Phys. Lett, A38, 339 (1972)
63. R. Vilanove, F. Rondelez, Phys. Rev. Lett. 45, 1502 (1980)
64. D.K. Chattoraj, K.S. Birdi, Adsorption and the Gibbs surface excess, Plenum Press, New York (1984)
65. A. Leiva, L. Gargallo, A. Gonzalez, D. Radić, Eur. Polym. J. 40, 2349 (2004)
66. S. Havlin, D. Ben Avraham, Phys. Rev. A 27, 2759 (1983)
67. R. Vilanove, D. Poupinet, F. Rondelez, Macromolecules, 19, 1054 (1986)
68. L. Gargallo, B. Miranda, A. Leiva, D. Radić, Polymer, 46, 824 (2001)
69. L. Gargallo, A. Leiva, M. Urzua, L. Alegria, B. Miranda, D. Radić, Polym. Int., 53, 1652 (2004)
70. L. Gargallo, B. Miranda, A. Leiva, D. Radić, M. Urzua, H. Rios, Polymer, 45, 5145 (2004)
71. I.I. Maleev, N.S. Tsvetkov, I.E. Twardon, Sintez Fiziko-Khim Polim. 16, 111 (1975)
72. P. Baglioni, E. Gallori, G. Gabrielli, C. Ferroni, J. Colloid Interface Sci., 88, 221 (1982)
73. J. Israelachvili, Langmuir, 10, 3774 (1994)
74. Z. Li, W. Zhao, J. Quinn, M.H. Rafailovich, J. Sokolov, R.B. Lennox, A. Eisenberg, X.Z. Wu, M.W. Kim, S.K. Sinha, M. Tolan, Langmuir, 11, 4785 (1995)
75. J. Zhu, A. Eisenberg, R.B. Lennox, Macromolecules, 25, 6547 (1992)
76. C. Chovino, P. Gramain, Macromolecules, 31, 7111 (1998)
77. J.R. Dann, J. Colloid Interface Sci., 32, 302 (1970)
78. P.G. de Gennes, Adv. Colloid Interface Sci., 27, 189 (1987)
79. M. Urzua, L. Gargallo, D. Radić, J. Macromol. Sci. Pure Appl. Chem., A37, 37 (2000)
80. L Gargallo, B. Miranda, H. Rios, F. Gonzalez – Nilo, D. Radić, Polym. Int., 50, 858 (2001)
81. L. Gargallo, B. Miranda, A. Leiva, H. Rios, F. Gonzalez-Nilo, D. Radić, J. Colloid Int. Sci., 271, 181 (2004)
82. M. Kawaguchi, S. Itoh, A. Takahashi, Macromolecules, 20, 1052 (1987)
83. M. Kawagushi, S. Itoh, A. Takahashi, Macromolecules, 21, 1056 (1987)
84. F. Davis, P. Hodge, X.-H. Liu, Z. Ali – Adib, Macromolecules, 27, 1957 (1994)
85. P. Dynarowicz-Latka, A. Dhanabalan, O. Oliveira, J. Phys. Chem. B, 103, 5992–6000 (1999)
86. N.R. Pallas, B.A. Pethica, Langmuir, 1, 509 (1995)
87. J.C. Earnshaw, P.J. Winch, Thin Solid Films, 159, 159 (1988)
88. D.M. Taylor, Y. Dong, C.C. Jones, Thin Solid Films 284, 5, 130 (1996)
89. J. Krägel, G. Kretzchmar, J.B. Li, G. Loglio, R. Miller, Thin Solid Films, 284–285, 361 (1996)
90. R.C.W. Liu, S. Cantin, F. Perrot, F.M. Winnik, Polym. Adv. Technol., 17, 798 (2006)
91. H.G. Schild, Prog. Polym. Sci., 17, 163 (1992)
92. X. Wang, X. Qiu, C. Wu, Macromolecules, 31, 2972 (1998)
93. W. Saito, M. Kawaguchi, T. Kato, T. Imae, Langmuir, 12, 5947 (1996)
94. P. Kujawa, C.C.E. Goh, D. Calvet, F.M. Winnik, Macromolecules, 34, 6387, (2001)
95. P. Kujawa, R.C.W. Liu, F.M. Winnik, J. Phys. Chem. B., 106, 5578 (2002)
96. B.A. Noskov, T.U. Zubkova, J. Colloid Interface Sci., 170, 1 (1995)
97. M. Kawaguchi, Prog. Polym. Sci., 18, 341 (1993)
98. F.C. Goodrich, Proc. R. Soc. London, Ser. A, 374, 341 (1981)
99. H.E. Gaub, H.M. McConnell, J. Phys. Chem., 90, 6830 (1986)
100. B.A. Noskov, Colloid Polym. Sci., 273, 263 (1995)
101. M. Kawaguchi, B.B. Sauer, H. Yu, Macromolecules, 22, 1735 (1989)
102. A. Halperin, M. Tirrell, T.P. Lodge, Adv. Polym. Sci., 100, 31 (1992)

103. K.R. Shull, E.J. Kramer, Macromolecules, 23, 4769 (1990)
104. Z. Tuzar, P. Kratochvil, Adv. Colloid Interfaces Sci., 6, 201 (1976)
105. A. Halperin, Macromolecules, 20, 2943 (1987)
106. A. Johner, J.F. Joanny, Macromolecules, 23, 5299 (1990)
107. J. Zhu, A. Eisenberg, R.B. Lennox, Macromolecules, 25, 6556 (1992)
108. M. Niwa, N. Katsurada, N. Higashi, Macromolecules, 21, 1878 (1988)
109. I. Langmuir, J. Chem. Phys. 1, 756, (1933)
110. S. Li, C.J. Clarke, R.B. Lennox, A. Eisenberg, Colloid Surf. A 133, 191 (1998)
111. Z. Li, W. Zhao, J. Quinn, M.H. Rafailovich, J. Sokolov, R.B. Lennox, A. Eisenberg, X.Z. Wu,
      M.W. Kim, S.K. Sinha, M. Tolan, Langmuir, 11, 4785 (1995)
112. K. Letchford, H. Burt, Eur. J. Pharm. Biopharm. 65, 259 (2007)
113. A. Halperin, J. Macromol. Sci., Part C: Polym. Rev., 46, 173 (2006)
114. O.V. Borisow, E.B. Zhulina, Macromolecules, 38, 2506 (2005)
115. P.G. de Gennes, Adv. Colloid Interface Sci., 27, 180 (1987)
116. C.M. Marques, J.F. Joanny, Macromolecules, 22, 1454 (1989)
117. J.A. Hubbell, Science 300, 595 (2003)
118. D.J. Pochan, Z. Chen, H. Cui, K. Hales, K. Oi, K.L. Wooley, Science, 306, 94 (2004)
119. R. Mezzenga, J. Ruokolainen, G.H. Fredrickson, E.J. Kramer, D. Moses, A.J. Heeger,
      O. Ikkala, Science, 299, 1872 (2003)
120. T.P. Russell, Science, 297, 964 (2002)
121. C.A. Devereaux, S.M. Baker, Macromolecules, 35, 1921 (2002)
122. Y. Seo, J.H. Im, J.S. Lee, J.H. Kim, Macromolecules, 34, 4842 (2001)
123. Y. Seo, A.R. Esker, D. Sohn, H.J. Kim, S. Paek, H. Yu, Langmuir, 19, 3313 (2003)
124. Y. Ikada, H. Iwata, S. Nagaoka, F. Horii, J. Macromol. Sci., Phys. B17, 191 (1980)
125. E. Bringuier, R. Vilanova, Y. Gallot, J. Selb, F. Rondelez, J. Colloid Interface Sci., 104, 95
      (1985)
126. M. Niwa, T. Hayashi, N. Higashi, Langmuir, 6, 263 (1990)
127. M. Yoshikawa, J.D. Worsfold, T. Matsuura, A. Kimura, T. Shimidzu, Polym. Commun. 31,
      414 (1990)
128. J. Kumaki, Macromolecules, 21, 749 (1988)
129. A. Wesemann, H. Ahrens, R. Steitz, S. Föster, C.A. Helm, Langmuir, 19, 709 (2003)
130. S. Yamamoto, Y. Tsujii, K. Yamada, T. Takeshi, T. Miyamoto, Langmuir, 12, 3671 (1996)
131. Y. Kim, J. Pyun, J.M.J. Frechet, C.J. Hawker, C.W. Frank, Langmuir, 21, 10444 (2005)
132. Y. Park, Y-W. Choi, S. Park, C.S. Cho, M.J. Fasolka, D. Sohn, J. Colloid Interf. Sci., 283,
      322 (2005)
133. C.S. Cho, J.W. Nah, Y.I. Jeong, J.B. Cheon, S. ASayama, H. Ise, T. Akaike, Polymer, 40,
      6769 (1999)
134. H. Yim, M.D. Foster, J. Engelking, H. Menzel, A.M. Ritcey, Langmuir, 16, 9792 (2000)
135. S.Garnier, A. Laschewsky, Langmuir, 22, 4044 (2006)
136. S.Creutz, J. van Stam, F.C. De Schryver, R. Jerome, Macromolecules, 31, 681 (1998)
137. G. Riess, Prog. Polym. Sci., 28, 1107 (2003)
138. J.B. Vieira, R.K. Thomas, Z.X. Li, J. Penfold, Langmuir, 21, 4441 (2005)
139. S. Antoun, J.F. Gohy, R. Jerome, Polymer, 42, 3641 (2001)
140. G. Riess, C. Labbe, Macromol. Rapid. Commun. 25, 401 (2004)
141. T. Rager, W.H. Meyer, G. Wegner, Macromolecules, 30, 4911 (1997)
142. J. Des Cloizeaux, J. Phys. France, 49, 699 (1988)
143. P.G. de Gennes, Macromolecules, 14, 1637 (1981)
144. G.J. Fleer, J.M.H.M. Scheutjens, M.A. Cohen, Stuart. Colloid Surf., 31, 1 (1988)
145. T.C. Patton, "Paint Flow and Pigment Dispersion", Interscience, New York (1964)
146. D.K. Owens, R.C. Wendt, J. Appl. Polym. Sci. 13, 1741 (1969)
147. D.H. Kaeble, J. Adhes., 2, 50 (1970)
148. R.C. Hayward, A.S. Utada, N. Dan, D.A. Weitz, Langmuir, 22, 4457 (2006)

Chapter 4
Supramolecular Structures: Complex Polymeric Systems – Organization, Design and Formation using Interfaces and Cyclic or Complex Molecules

Summary The analysis of supramolecular structures containing polymers, and the discussion about the effect of polymeric materials with different chemical structures that form inclusion complexes is extensively studied. The effect of the inclusion complexes at the air–water interface is discussed in terms on the nature of the interaction. The entropic or enthalpic nature of the interaction is analyzed. The description of these inclusion complexes with different cyclodextrines with several polymers is an interesting way to understand some non-covalent interaction in these systems. The discussion about the generation and effect of supramolecular structures on molecular assembly and auto-organization processes is also presented in a single form. The use of block copolymers and dendronized polymers at interfaces is a new aspect to be taken into account from both basic and technological interest. The effect of the chemical structure on the self-assembled systems is discussed.

Keywords Supramolecular system · Block copolymer · Self-assembled system · Inclusion complexes · Cyclodextrin · Air-water interface · Non-covalent interaction · Auto-organization

4.1 Introduction

The importance of non – covalent interactions in biological systems has motivated much of the current interest in supramolecular assemblies [1]. A classical example of a supermolecule has been provided by the rotaxanes [2, 3], in which a molecular "rotor" is threaded by a threaded by a linear "axle". Another examples have been previously included as cyclic crown ethers threaded by polymers, paraquat – hydroquinone complexes [4] and cyclodextrin complexes [5, 6].

The generation of supramolecular structure of interest usually relies on molecular self-assembly and auto-organization processes. Much attention has been recently focused on the design of nanometer-scale (nanoscale) molecular devices. One approach to the molecular devices is the self – assembly of supramolecular structures such as the inclusion complexes [1, 2, 7, 8].

L. Gargallo, D. Radić, *Physicochemical Behavior and Supramolecular Organization of Polymers*, DOI 10.1007/978-1-4020-9372-2_4,
© Springer Science+Business Media B.V. 2009

Recently, increasing attention has been paid to polymer inclusion complexes (ICs) formed between various polymers and cyclodextrins (CDs) or an other complex molecules. Such ICs are based on noncovalent host – guest interactions and useful building blocks for constructing supramolecular structures [3–5, 9–11]. For a long time, it was known that α-, β- and γ – cyclodextrins (CDs) consisting of six, seven and eight glucopyranose units, respectively, form complexes with several kinds of organic molecules, incorporating them as guests within their cavity. Figure 4.1 shows the chemical structure and dimensions of the α-, β – and γ – cyclodextrins [12].

The most important characteristics of the CDs are summarized in Table 4.1.

Cyclodextrins comprises a family of three well-known industrially produced major, and several rare, minor cyclic oligosaccharides. The three major CDs are crystalline, homogeneous, nonhygroscopic substances, which are torus-like macro-rings built up from glucopyranose units [13].

As a consequence of the $^4C_1$ conformation of the glucopyranose units, all secondary hydroxyl groups are situated on one of the two edges of the ring, whereas all the primary ones are placed on the other edge. The ring, in reality, is a cylinder, or better said a conical cylinder, which is frequently characterized as a doughnut or wreath-shaped truncated cone. The cavity is lined by the hydrogen atoms and the glycosidic oxygen bridges, respectively. The nonbonding electron pairs of the glycosidic-oxygen bridges are directed toward the inside of the cavity, producing a high electron density there and lending to it some Lewis-base character.

The C-2-OH group of one glucopyranoside unit can form a hydrogen bond with the C-3-OH group of the adjacent glucopyranose unit. In the βCD molecule, a complete secondary belt is formed by these H bonds, therefore, the βCD is a rather rigid structure. This intramolecular H-bond formation is probably the explanation for the observation that βCD has the lowest water solubility of all CDs.

The H-bond belt is incomplete in the αCD molecule, because one glucopyranose unit is in a distorted position. Consequently, instead of the six possible H bonds, only four can be established simultaneously. The γCD is a non-coplanar, more flexible structure; therefore, it is the more soluble of the three CDs.

Figure 4.2 shows a sketch of the characteristic structural features of CDs. On the side where the secondary hydroxyl groups are situated, the cavity is wider than on the other side where free rotation of the primary hydroxyls reduces the effective diameter of the cavity.

Cyclodextrins are really, the most widely used molecules that form host/guest-type inclusion complexes [13].

Potential use of complexes of γ – CD with organic compounds, including polymers, was also reviewed by Szejtli [6, 12, 13]. γ – CD are able to incorporate metal ions as ligands to prepare magnetic nano – particles [7, 14]. Harada and Kamachi [8, 15] first found that poly (ethylene oxide) (PEO) thread α– CD rings to form polymer – cyclodextrin complex. Since their finding of inclusion complex formation of polymer chains with α– CD, a large number of studies on inclusion complexes of

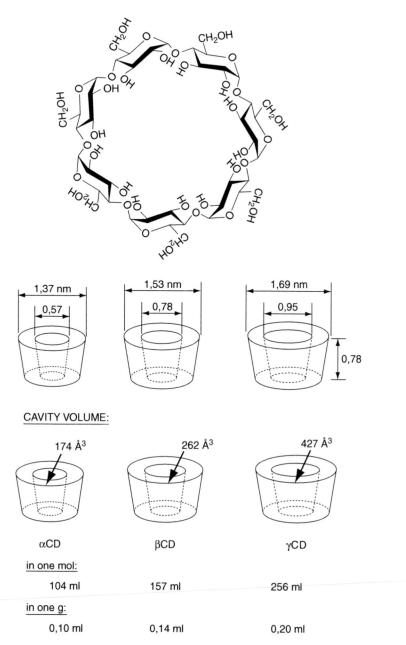

CAVITY VOLUME:

174 Å³	262 Å³	427 Å³

αCD   βCD   γCD

in one mol:

104 ml   157 ml   256 ml

in one g:

0,10 ml   0,14 ml   0,20 ml

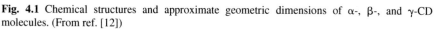

Fig. 4.1 Chemical structures and approximate geometric dimensions of α-, β-, and γ-CD molecules. (From ref. [12])

Table 4.1 Characteristics of α, β, and γ CDs. (From ref. [12])

	α	β	γ
no. of glucose units	6	7	8
mol wt	972	1135	1297
solubility in water. g $100\,mL^{-1}$ at room temp	14.5	1.85	23.2
$[\alpha]_D 25°C$	150 ± 0.5	162.5 ± 0.5	177.4 ± .5
cavity diameter, Å	4.7 − 5.3	6.0 − 6.5	7.5 − 8.3
height of torus, Å	7.9 ± 0.1	7.9 ± 0.1	7.9 ± 0.1
diameter of outher periphery, Å	14.6 ± 0.4	15.4 ± 0.4	17.5 ± 0.4
approx volume of cavity, $Å^3$	174	262	427
approx cavity volume in 1 mol CD (ml)	104	157	256
in 1 g CD (ml)	0.10	0.14	0.20
crystal forms (from water)	hexagonal plates	monoclinic parallelograms	quadratic prisms
crystal water, wt%	10.2	13.2 − 14.5	8.13 − 17.7
diffusion constant at 40°C	3.443	3.224	3.000
hydrolysis by *A. oryzae* α-amylase	negligible	slow	rapid
V_{max} value, min^{-1}	5.8	166	2300
relative permittivity (on incorporating the toluidinyl group of 6-*p*-toluidynilnaphthalene 2-sulfonate) at pH = 5.3, 25°C	47.5	52.0	70.0
(on incorporating the naphthalene group)	a	29.5	39.5
pK (by potentiometry) at 25°C	12.332	12.202	12.081
partial molar volumes in solution mL mol^{-1}	611.4	703.8	801.2
adiabatic compressibility in aqueous solutions mL $(mol^{-1}\,bar^{-1}) \times 10^4$	7.2	0.4	−5.0

^{a} Naphthalene group is too bulky for the α-CD cavity.

α -, β - and γ - CDs with various polymers have been reported [9,16], particularly by Harada's group covering hydrophilic and hydrophobic polymers [10,11,15,17–22].

The crystalline structures of inclusion complexes of γ-cyclodextrin (γ-CD) with poly (ethylene glycol), poly (ethylene adipate), poly (propylene glycol) and poly (isobutylene) were studied by electron microscopy, in combination with X-ray diffraction works and measurements of thermal properties by DSC and TGA [16]. It was found that water molecules were inevitable to form crystalline inclusion complexes of γ-CD with the polymers. There are three modifications depending on the content of water in the complexes. Which crystalline form is stable is dependent on the content of water hydrates in the complexes [16].

New supramolecular assemblies based on a chitosan bearing pendant cyclodextrins were prepared [23]. In order to obtain these assemblies, adamantyl groups which can selectively be included in the cyclodextrins cavity were grafted on chitosan and several poly(ethylene glycol)s (PEG) with molecular weights of 3400,

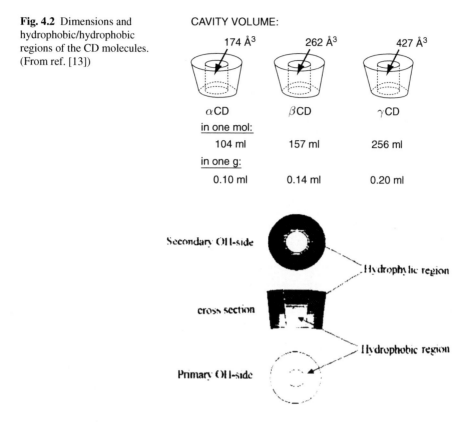

Fig. 4.2 Dimensions and hydrophobic/hydrophobic regions of the CD molecules. (From ref. [13])

6000, and 20000, affording guest macromolecules, with different structural features. The stoichiometry of the complex of PEG – Diadamantane with β – CD was found to be 1:1 which, according with Auzely-Velty and Rinaudo [23], would imply that the inclusion complex can be considered as a first – order system.

The average apparent association constant K_a value that was derived from a numerical simulation of the experimental data [24] was found to be similar to that found for inclusion of 1 – adamantanecarboxylate in β – CD [25]. The values for K_a at 35, 45, and 55°C were also determined in order to estimate the thermodynamic parameters for inclusion complexation. Table 4.2 summarized the results.

Table 4.2 Thermodynamic Parameters for Inclusion Complex Formation of PEG – Diadamantane with β-CD. (From ref. [23])

T (°C)	K_a (M^{-1})	$\Delta G°$ (kJ/mol)	$\Delta H°$ (kJ/mol)	$T\Delta S°$ (kJ/mol)
25	$(1.8 \pm 0.09) \times 10^4$	-24.3 ± 0.1	-25.5 ± 0.5	1.1 ± 1
35	$(1.37 \pm 0.07) \times 10^4$	-24.4 ± 0.1		
45	$(9.98 \pm 0.5) \times 10^3$	-24.3 ± 0.1		
55	$(7 \pm 0.4) \times 10^3$	-24.1 ± 0.1		

A large negative ΔH and a near zero ΔS were found for the complex. This result would suggest that inclusion of the adamantyl moiety in β – CD is an enthalpy-driven process and that K_a decreases when the temperature increases.

These data thus provide evidence of specific interactions between grafted adamantane groups and β- CD with no influence of the polyether chain in the inclusion process [23]. And at the same time was confirmed by ^{1}H NMR spectra the selective interaction between the pendant hydrophobic adamantyl groups and β – cyclodextrin.

4.2 Inclusion Complexes Between Polymers and Cyclic Molecules Surface Activity

Although a wide range of polymers have been investigated with various cyclodextrins, these studies mainly focused on the IC preparation techniques and characterization of solid phases. The solution properties, such as the self – assembly behavior, dissociation, particle size an surface activity, were not commonly reported. These solution properties, especially the assembly and surface behavior, are vital for the potential applications of such systems in biomedical science, such as in controlled drug delivery.

In an aqueous solution, the slightly apolar cyclodextrin cavity is occupied by water molecules that are energetically unfavored (polar-apolar interaction), and therefore can be readily substituted by appropriate "guest molecules", which are less polar than water (Fig. 4.3). The dissolved cyclodextrin is the "host" molecule, and part of the "driving force" of the complex formation is the substitution of the high-enthalpy water molecules by an appropriate "guest" molecule. One, two or three CD molecules.

Mixtures of poly (ethylene oxide) (PEO) of various molecular weights with α – CD have given stoichiometric complexes in high yields [17]. It is important to consider that the formation of the complexes involved the threading of the α-CD along the polymer chain into a "necklace-like" structure [26]. This process is driven

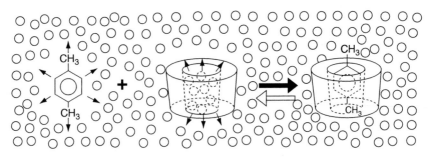

Fig. 4.3 Schematic representation of the association-dissolation of the host (cyclodextrin) and guest (p-xylene). The formed guest/host inclusion complex can be isolated as a microcrystalline powder. (From ref. [13])

by noncovalent attractive forces, therefore allowing the α-CD to slide along the polymer backbone [27]. The formation of the supramolecular adduct is entropically unfavorable for the guest polymer, as the linear polymer chain must fit into several host units to produce the final complex. The complex is thought to be promoved by hydrophobic interactions between the cavity of α-CD and the -CH2CH2O- units of PEO and also by hydrogen – bond formation between the hydroxyl groups situated along the rim of α-CD molecules threaded onto the PEO chain [27,28]. This peculiar organization should present behavior different from that of α-CD and PEO at the air – aqueous interface.

Recently the surface properties of the supramolecular inclusion complex (ICs) obtained from the threading of α-CD onto poly(ethylene oxide) (PEO) free in solution was studied [28]. The complex were characterized by IR, ^{1}H NMR spectroscopy, and thermal analisis. The variation of the interfacial tension, γ_{int}, with inclusion complex (IC) concentration and temperature were determined. The results were compared with those found for PEO under the same conditions. α-CD does not present surface activity [28]. To quantify the adsorption process of IC and PEO in aqueous medium, the following form of Gibbs equation was used [29].

$$\Gamma = -(RT) - 1Cp(d\gamma_{int}/d\ Cp) \qquad (4.1)$$

Where Γ is the excess surface concentration and R and T have their usual meanings. In order to evaluate the slopes, $d\gamma_{int}/dCp$, the experimental data of $d\gamma_{int}$ versus C_{IC} and C_{PEO} can be adjusted to the empirical equation of Szyszkowski [30].

$$\gamma_{int} = \gamma_{int}^o - \gamma_{int}B\ log[(C_{IC}/A) + 1] \qquad (4.2)$$

where γ^O int is the interfacial tension between the pure phases, and A and B values are two numerically empirical fitting parameters. By iteration, the best A and B values were obtained. By differentiation of this equation it was possible to obtain.

$$d\gamma_{int}/dC_{IC} = -\gamma_{int}^o B/(C_{IC} + A) \qquad (4.3)$$

which can be related to equation (4.1) to give the Gibbs – Szyskowski equation. Using this combined equation it was possible to determined Γ^∞, the limiting excess surface concentration. The Γ^∞ value is related to the area covered by an average monomer unit, according to

$$\sigma = (\Gamma^\infty N) - 1 \qquad (4.4)$$

Where N is the Avogadro constant.

Figures 4.4a and b show the adsorption process of IC in the high – dilution zone at different temperatures (283, 293, 298, 303, and 308 K ($\pm$ 0.1 K).

As can be seen, γ_{int} decreases with IC concentration until a plateau seems to be reached. γ_{int} is clearly depending on the temperature for this system. The slope $d\gamma_{int}/dC_{IC}$ is more negative as the temperature decreases. Different behavior was

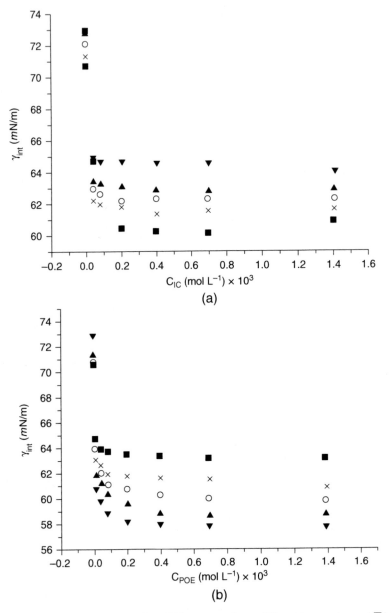

Fig. 4.4 Interfacial tension as a function of IC concentration at different temperatures : (■) 288; (x) 293; (o) 298; (▲) 303; (▼) 308 K. (From ref. [28])

Table 4.3 Thermodynamic parameters for the adsorption of IC and PEO indilute aqueous solutions. (From ref. [28])

System	ΔG^0_{ads} (kJ mol^{-1})	ΔH^0_{ads} (kJ mol^{-1})	$T\Delta S_{ads}$ (kJ mol^{-1})
IC	−30.8	−92.5	−61.7
PEO	−30.4	37.5	67.9

found in aqueous solutions of PEO in the same range of concentration. In this case, γ int decreases with PEO concentration and $d\gamma$ int/dC_{PEO} is more negative as the temperature increases.

The thermodynamic parameters for the adsorption process of the IC and PEO were also determined using the classical equation [31]. The thermodynamic parameters for the adsorption process of IC are listed in Table 4.3.

The analysis of these experimental data, taken from reference [28], shows that the thermodynamic parameter values of IC and PEO are completely different. A large positive ΔH^O and also a large positive ΔS^O are found for the adsorption process of PEO, suggesting that the driving force for this process is of an entropic nature and then ΔG^O is more negative when the temperature increases.

Figure 4.5a and b show the thermodynamical behavior observed for these systems.

Inclusion complex formation between polyaniline with emeraldine base and β-cyclodextrin has been studied by the frequency – domain electric birefringence (FEB) spectroscopy in a solution of N-methyl-2-pyrrolidone (NMP) and by scanning tunneling microscopy (STM). The FEB results show that polyaniline in the solution with cyclodextrin changes its conformation from coil to rod at low temperature (below 275 K), and some rodlike images are observed on a substrate by STM. These results have suggested that cyclodextrins are threaded onto polyaniline and confine the conformation of the polymer chain to a rodlike one. Furthermore, it is found that the threaded cyclodextrins prevent the chemical oxidation, i.e., doping of polyaniline by iodine. This indicates formation of a new inclusion complex, a conjugated conducting polymer covered by insulated cyclic molecules, namely, "insulated molecular wire".

Figure 4.6 shows the schematic diagrams of ciclodextrins, polyaniline with emeraldine base, and inclusion complex formation of cyclodextrins and a conducting polymer chain: insulated molecular wire.

These results have been taken from the Yoshida et al. [32]. The same authors have studied the inclusion complexes between these cyclic molecules and conjugated conducting polymers using FEB and STM microscopy. Since the FEB signal, or the Kerr effect, comes from optical and electrical anisotropy of molecules, rodlike molecules such as liquid crystals, tobacco mosaic virus, polypeptides, and linear polyions yield large electric birefringence but isotropoic molecules such as coiled polymer chains and spherical lattices exhibit no signal [18,33]. By this way the FEB technique is a useful tool to determine whether the conformation of a polymer chain is rodlike or coiled in solution. Shimomura et al. [34] have also investigated the rod-coil transition of a conjugated conducting polymer in solution by FEB.

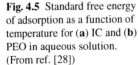

Fig. 4.5 Standard free energy of adsorption as a function of temperature for (**a**) IC and (**b**) PEO in aqueous solution. (From ref. [28])

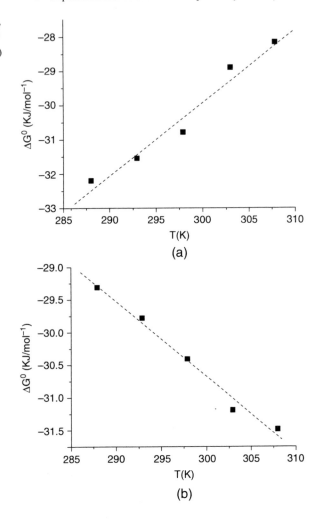

The studies reported in the literature about the inclusion complexes between polymers and cyclodextrins have shown a good correlation between the cross-sectional areas of the polymers and the cavity size of the CDs [3, 9]. For example, poly(ethylene glycol) (PEG) can form inclusion complexes with α-CD [8, 15, 20, 35] and γ-CD [36, 37] while poly(propylene glycol) (PPG) can form inclusion complexes with β-CD [23, 24, 38, 39]. Except for a few cases where block polymers were used, most studies dealt with polymers having "homogeneous" cross-sectional areas along the polymer chains. The behavior of polymer IC formation for a polymer with "heterogeneous" cross-sectional area as poly(isobutylene) (PIB) was studied by Jiao et al. [40]. In this case, the polymer does not form complexes with α-CD at any molecular weight because of the hindrance exerted by dimethyl groups on the main chain [41].

Fig. 4.6 Schematic diagrams of (**a**) cyclodextrins, (**b**) poly-aniline with emeraldine base, and (**c**) inclusion complex formation of cyclodextrins and a conducting polymer chain: insulated molecular wire. (From ref. [32])

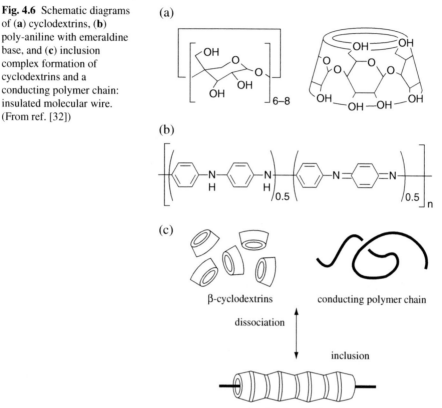

Poly (ε- caprolactone) (PEC) forms inclusion complexes with all three CDs wherein the γ-CD/PCL complex contains two side-by-side PCL chains in each γ-CD channel when the molecular weight of PCL is low [26, 42].

The proposed structures by Kawaguchi et al. [42] for these complexes are shown in Fig. 4.7.

Jiao [40] has reported results of Poly (neopentyl glycol diacid ester)s (PNEs) which present an intermediate structures between those of PIB and PCL. Scheme 4.1 shows the structures of PIB, PCL, and PNE.

Varias structures can be obtained by changing the length of the diacid monomer, and such structural change modifies the polymer IC formation behavior. The inclusion complexes of poly(neopentyl glycol sebacate) (m = 8, in Scheme 4.1) (PNGS) with CDs. PNGS was able to form inclusion complexes with all three CDs in moderate yields. For the α-CD/PNGS complex, it was suggested that the conformational flexibility of both α-CD and PNGS enables α-CD to squeeze over the bulky dimethyl groups and settle on and complex the thinner part of the polymer chains [25, 40].

Fig. 4.7 Proposed structure
of the α-CD-PEC and
γ-CD-PEC complexes. (From
ref. [42])

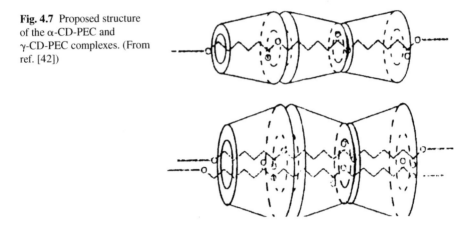

The driving forces of complex formation were thought to be the geometric com-
patibility or fit and intermolecular interaction between hosts and guests. It has been
reported that many linear polymeric guests could form inclusion complexes with
CDs resulting in main-chain pseudopolyrotaxanes. When the polymers were added
into the CD solutions and then sonicated, crystalline inclusion complexes precipi-
tated. As the result of X- ray diffraction study, all crystalline inclusion complexes
between CDs and polymeric guests are columnar in structure [27,43].

As Tonelli et al. [44,45] have pointed out, the study of crystalline inclusion complexes provides an approach to investigate the behaviors of single polymer chains
in isolated and well – defined environments. Then, it is helpful in understanding the
mechanism of molecular recognition between hosts and polymeric guests.

It has been also reported a study on the threading process of a α – cyclodextrin
(α-CD) and polyethylene glycol (PEG), as a function of temperature and solvent
composition. This reaction produces a polyrotaxane that eventually precipitates and
forms a thick gel. Ceccato et al. [46] have proposed a molecular model for the interpretation of the temperature and solvent composition effect on the threading process.
According of this model, the reaction can be depicted as a five – step phenomenon
that mainly depend on the threading and sliding of α-CD and PEG. The transition

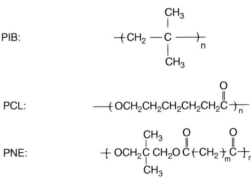

Scheme 4.1 (From ref. [40])

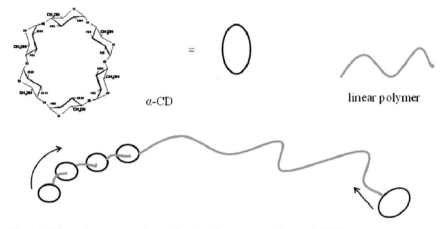

α-CD

linear polymer

Fig. 4.8 Schematic representation of the threading process. (From ref. [46])

state theory provides a way to calculate the Gibbs free energy change related to this process and the number of α-cyclodextrin molecules that participate in the formation of the polyrotaxane. This number was in good agreement with the value predicted according to the literature and geometrical considerations. $\Delta G^{\ddagger}$ parameters depends on the nature of the solvent and was related to the interactions of PEG and α-CD with the solvent molecules.

Figure 4.8, shows a squematic representation of the threading process, taken from [46]

The selectivity of the Complex Formation is a very interesting subject. γ – Cyclodextrin, (γ – CD) has been found to form inclusion complexes with poly (methyl vinyl ether) (PMVE), poly(ethyl vinyl ether) (PEVE), and poly(n- propyl vinyl ether) (PnPVE) of various molecular weights to give stoichiometric compounds in crystalline states. However, α- cyclodextrin (α – CD) and β – Cyclodextrin (β- CD) did not form complexes with poly (alkyl vinyl ether)s of any molecular weight. γ –CD did not form complexes with the low molecular weight analogs, such as diethyl ether and trimethylene glycol dimethyl ether.

Scheme 4.2 shows the chemical structures of the poly (alkyl vinyl ether)s [47] reported.

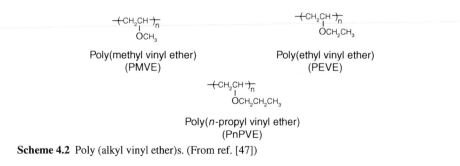

Poly(methyl vinyl ether)
(PMVE)

Poly(ethyl vinyl ether)
(PEVE)

Poly(n-propyl vinyl ether)
(PnPVE)

Scheme 4.2 Poly (alkyl vinyl ether)s. (From ref. [47])

Molecular model studies have shown that PMVE, PEVE, and PnPVE chains are capable to penetrate γ –CD cavities. Model studies further indicate that the single cavity can accommodate three monomer units. The inclusion complex formation of polymers with cyclodextrins is entropically unfavorable. However, formation of the complexes is thought to be promoted by hydrogen bond formation between cyclodextrins. Therefore, the head – to – head and tail – to – tail arrangement, which results in a more effective formation of hydrogen bonds between cyclodextrins, is thought to be the most probable structure. This structure was proved by X-ray studies on a single crystal of the complex between γ -CD and 1-propanol.

Figure 4.9 taken from [47] shows a proposed structure of the complex between γ-CD and PMVE.

More recently, it has been reported [48] an attempt to blend polymers by first forming their common inclusion compound (IC) with cyclodextrins (CD) as the host and then coalescing the guest polymer from their CD – IC crystals by washing them with hot water. This procedure was used in the hope of obtaining an intimately mixed, compatible blend of the poly (ε –caprolactone) (PEC)/poly (L-lactic acid) (PLLA) pair, which are normally incompatible. Tonelli et al. [48, 49] have reported observations made on the poly(carbonate) (PC)/poly(methyl methacrylate) (PMMA) pair, which are respectively difficult to crystallize and amorphous. PC/PMMA blends coalesced from their common γ – CD – ICs are amorphous and generally exhibit single glass transitions at temperatures (Tg) between those of pure PC and PMMA. But, at a 1:4 molar PC:PMMA blend coalesced from its common γ – CD – IC is characterized by a Tg lower than that of pure PMMA. FTIR spectroscopy has suggested an intimate mixing of and possible specific interactions between PC and PMMA chains in the coalesced blends as reflected by substantial shifts in the frequencies of the PMMA and PC C = O vibrations.

In conclusion, it has found, when inherently immiscible polymers are simultaneously included as guests in the narrow channels of their common inclusion compounds (ICs) formed with host cyclodextrins (CDs) and then these polymer-1/polymer-2-CD-IC crystals are washed with hot water to remove the host CD lattice

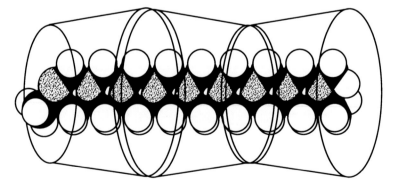

Fig. 4.9 Proposed structure of the complex between γ-CD and PMVE. (From ref. [47])

and coalesce the guest polymers, that an intimately mixed blend of the polymers are obtained.

In another study, it was successfully reported an intimate ternary blend system of poly(carbonate) (PC)/poly(methyl methacrylate) (PMMA)/poly (vinyl acetate) (PVAc) obtained by the simultaneous coalescence of the three guest polymers from their common γ-cyclodextrin (γ-CD) inclusion complex (IC). The thermal transitions and the homogeneity of the coalesced ternary blend were studied by differential scanning calorimetry (DSC) and thermogravimetric analysis (TGA) [50]

Recently, increasing attention has been paid to polymer inclusion complexes (PICs) formed between various polymers and cyclodextrins (CDs). Such PICs are based on noncovalent host – guest interaction and are useful building blocks for constructing supramolecular structures [9, 10]. Lui et al. [51] have reported a PIC produced from α-CD and a double – hydrophilic diblock copolymer, poly(ethylene oxide) –b- poly(acrylic acid) (PEO-b-PAA), which focused on the solution property and the mechanism driving the formation of the PIC. Complexes produced from a diblock copolymer and cyclodextrin were previously reported by Yui et al. [52] but their PIC product was gel-like. Double – hydroplilic poly (ethylene oxide)-b-Poly(acrylic acid) (PEO-b-PAA) self-assemble into nanostructures in basic solution upon the addition of α-CD as a result of the complexation between α-CD and PEO. The nanostructures produced were spherical in shape as was observed by transmission electron microscopy (TEM) and with a radii that were much larger than that a single stretched polymeric chain. The results have suggested that the nanostructures formed in the PEO-b-PAA/α-CD solution at high pH were likely to be spherical vesicules. Scheme 4.3 shows this particular behavior.

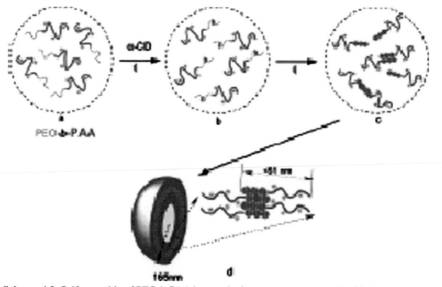

Scheme 4.3 Self-assembly of PEO-b-PAA into vesicular nanostructures at pH 10 induced by complexation between PEO segments and α–CD. (From ref. [51])

Girardeau et al. [53] have described the chain dynamics of PEO within nan-otubes of α-cyclodextrin using dueterated PEO (d-PEO) and ^{2}H solid – state NMR spectroscopy. The chain dynamics were explored and compared with the respective unthreaded d-PEO. As these materials are continually proposed for applications in molecular-level devices [54,55] characterization of their molecular dynamics is important since they play a key role in governing bulk physical properties.

From the studies of Girardeau et al. [53] it was possible to conclude that poly (ethylene oxide) chains isolated in cyclodextrin nanotubes exhibit faster dynamics as compared to bulk PEO at the same temperatures. In addition, PEO motions in the nanotubes are anisotropic and remain so even at temperatures above the melting point of uncomplexed PEO where dynamics are isotropic ($>$ 320 K). The motional geometry and activation energy from ^{2}H NMR spectra are consistent with trans – gauche conformational transitions, envisioned as gauche defects traveling along the mostly trans PEO chains within the nanotubes. The overall picture emerging from these dynamics studies is consistent with the morphology revealed by scattering experiments [56]. Longer CD nanotubes are formed by complexation with low-molecular-weight polymers, while high-molecular – weight polymers gel and lead to shorter CD nanotubes. As the length of the CD nanotube is decreased, the anisotropic motion occurs with much less defined jump angles and the ratio of unthreaded to threaded chain segments appears higher than expected given the cyclodextrin threading level. This was explained by increased exchange between threaded and unthreaded chain segments when the nanotubes are shorter.

Crystalline inclusion complexes (IC's) have been also formed between polymers and another small-molecules, host clathrated provide a unique environment for observing the solid – state behavior of isolated polymer chains. In their IC's with small-molecule, host clathrates, such as urea (U) [1] and perhydrotriphenylene (PHTP) [57], the included polymer chains are confined to occupy narrow channels (ca. 5.4 Å in diameter) where they are extended and separated from neighboring chains by the channel walls, which are composed exclusively of the host clathrate, crystalline matrix. Choi et al. [58] have been studied the behavior of isolated, extended polymer chains included in their IC's with U and PHTP by a combination of molecular modeling [59,60] and experimental observations in an effort to determine their conformations and mobilities in these well-defined, containing environments.

Molecular modeling of aliphatic polyesters and polyamides suggested [61] that both classes of polymers may be capable of forming these IC's. For example, it was suggested that poly(ε- caprolactone) (PEC) chains in either the all – trans or kink (g + tg+) conformations are slim enough to fit in these narrow IC channels (D = 5.5 A). Preliminary studies of its stability, stoichiometry, and structure, both the three – dimensional, solid – state structure of the PEC – U- IC and the conformation adopted by the included PEC chains have been reported [58].

The formation of the inclusion complexes has leaded to significant changes of the solubility and reactivity of the guest molecules, but without any chemical modification. Thus, water insoluble molecules may become completely water soluble simply by mixing with an aqueous solutions of CD of CD-derivatives. Based on this knowledge, Ritter et al. [62] were encouraged to investigate the behavior of CD-complexes of various monomers, e.g. of methacrylates or methacrylamides. The complexes monomers could be successfully polymerized via free radicals in water.

These investigations have demonstrated the successful application of cyclodextrins in polymer synthesis in aqueous solutions via free radical polymerization or via a oxidative recombination mechanism. Some special aspects of cyclodextrins were found concerning the kinetics, chain transfer reaction, and copolymerization parameters [63].

Due also to their controllable size, low cytotoxicity, and unique architecture, cyclodextrin-based polyrotaxanes and pseudopolyrotaxanes have been developed to encompass a broad range of diverse medical applications from erodable hydrogels to drug and gene delivery. A recent review about biomedical applications of cyclodextrin based polyrotaxanes have been reported by Loethen et al. [64].

This review has been focused on the literature relevant to pseudorotaxanes, pseudopolyrotaxanes, polyrotaxanes and rotaxanes that may potentially be used for a variety of drug delivery and medical imaging applications.

4.3 Self-Assemblies, Block Copolymers and Dendronized Polymers at the Interfaces: Effect of Molecular Architecture

There is a growing interest in polymers with architectures that differ from the classical linear polymers, since new polymer architectures may exhibit unusual behavior. For example, the combination of dendrimers and linears polymers in hybrid materials has evolved from a curiosity into an important trend in current chemistry. Therefore, various dendrimer construction strategies have been developed on the basis of classical organic and inorganic chemistry [65,66].

The development of well defined molecular and supramolecular architectures has attracted a strong scientific interest and the dendrimers also are among the most exciting molecular architectures.

Dendronized polymers are a class of polymers produced by the combination of linear polymers and dendritic molecules as side chain pendant moieties [67–69].

Scheme 4.4 shows one illustration of this particular structure taken from reference [69].

When dendritic fragments are attached to polymer chains, the conformation of the polymer chain is strongly affected by the size and chemical structure of the dendritic wedges attached. Dense attachment of dendritic side chain converts a linear polymer into a cylindrically shaped, rigid and nanoscopic dimension. Frechet and Hawker [70] were one of the first to recognize these "hybrid architectures".

Another property of this class of polymers is that the combination of specific dendrons with linear chains provides an opportunity to design a well-defined amphiphilic dendronized polymer system, which can bring about supramolecular aggregates in an aqueous phase [70,71].

It is possible to consider the dendrimers as branched macromolecules with a globular shape deviating significantly from that of linear coil polymers. Recently,

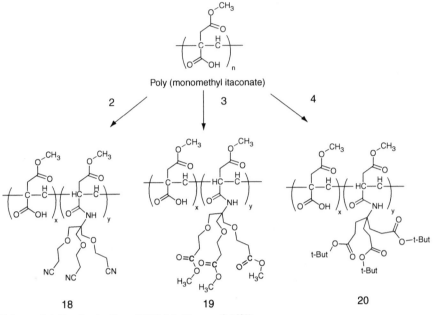

Scheme 4.4 Dendronization of PMMeI. (From ref. [69])

properties of block copolymer systems with a dendrimer and a linear coil block have been explored in solution, bulk and in thin – films [72]. Amphiphilic dendron – rod molecules with three hydrophilic poly (ethylene oxide) (PEO) branches attached to a hydrophobic octa – p – phenylene rod were also investigated for their ability to form two – dimensional micellar structures on a solid surface by J. Hozmueller et al. [73]. A treelike shape of the molecules was reported to be a mayor factor in the formation of nonplanar micellar structures in solution and in the bulk state. Hozmueller et al. have observed that in these treelike amphiphilic molecules the hydrophilic termi-nated dendron branches assemble themselves in surface monolayers with the forma-tion of two – dimensional layered or circular micellar structures.Similar molecules have been observed to organize into spherical aggregates in solution, but their ability to assemble into organized structures at the air – water interface have been only reported by J. Hozmueller et al. [73].

The rigid, hydrophobic rod core for both molecules was paired with three branched, flexible, hydrophilic PEO chains that possess excellent amphiphilic prop-erties. Scheme 4.5 shows the chemical structures of the complex molecules studied.

Three tetrabranched PEO chains were attached asymmetrically to a rigid octa – p –phenylene chain at the first and second phenyl rings as is shown in Fig. 4.10. The end functionality of the flexible PEO chains has been varied from methyl groups (molecule 1) to hydroxyl groups (molecule 2). Both molecules have displayed stable amphiphilic behavior at the air – water interface.

Figure 4.11 shows the surface – pressure isotherms obtained in both cases.

Molecule 1 underwent several phase transitions observed as the multiple shoul-ders and plateau regions in the pressure versus molecular area (π – A) isotherms.

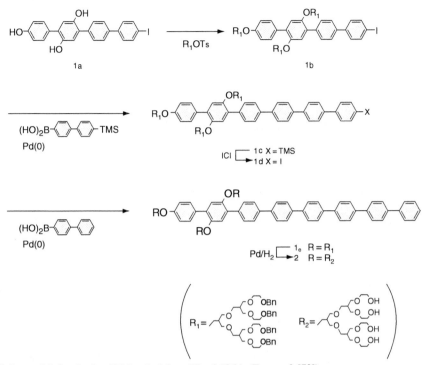

Scheme 4.5 Synthesis of Molecule 2 from Fig. 4.10(b). (From ref. [73])

The results have been discussed in terms of the chemical structure of the molecules studied and their surface morphology found.

In the area of supramolecular chemistry there are several aspects which must be studied. It is generally accepted that the functions of the polymeric materials, particularly, the supramolecular polymeric systems, are determined not only by the macromolecules themselves but also by how the macromolecules are arranged. Thus, the design of the macromolecules with different and complex structures as, block copolymers, polymercomplexes, inclusion complexes and supramolecular assemblies are very important. In the first case, one of the most fascinating properties of diblock copolymers are their ability to self – assemble into micelles, aggregates, and vesicles of several morphologies in the presence of a selective solvent [74–76]. It has been observed that the insoluble segment can organize into a core surrounded by the soluble segment as a corona to stabilize the aggregates. Depending upon polymer structure, composition and assembling conditions, block copolymers may self – organize into versatile particles, such as spheres, [77, 78] vesicles, [79] worms [80] and other complex assemblies [81]. Recent studies have demonstrated that self-assembly of diblock copolymers into various morphologies can occur not only in selective solvents but also at interfaces and surfaces [82, 83].

One of the first step toward understanding the surface behavior of these systems is to check the monolayer formation at the air – water interface. A system as an amphiphilic diblock copolymer for example, from the initial studies of them,

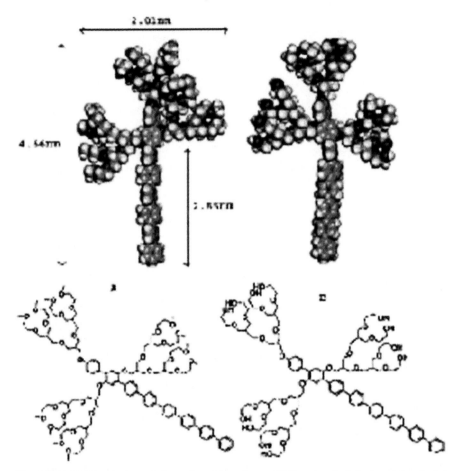

Fig. 4.10 (**a**) Molecule 1, methyl-terminated branches and (**b**) molecule 2, hydroxyl-terminated branches. Chemical structures are accompanied with molecular models. (From ref. [73])

it was shown that not in situ evidence of surface aggregation. The problem is to demonstrate if the diblock copolymers exist a individual molecules or as surface aggregates or supramolecular assemblies at the air – water interface. In the particular case of these amphiphilic diblock copolymers, it was possible to demonstrate that the hydrophilic segment could be the responsible of the surface behavior at the air – water interface irrespective of the nature of the other segment. Direct visualization of dramatic changes in the monolayer arrangement frequently involved the use of Brewster Angle Microscopy (BAM). The resolution of this technique is very important together with the Elipsometry.

In the supramolecular systems as the complexes of polymers and copolymers with cyclic molecules as cyclodextrins, the surface properties and the molecular motion must change when the polymer is free that when the polymer is included inside of a cyclic molecule. The polymer in the complexed form cannot have the same situation when it is in uncomplexed state.

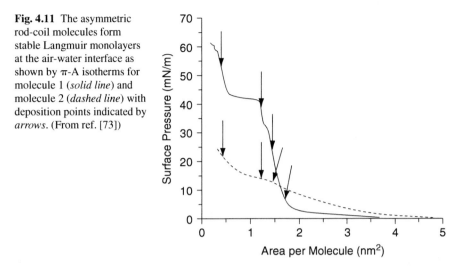

Fig. 4.11 The asymmetric rod-coil molecules form stable Langmuir monolayers at the air-water interface as shown by π-A isotherms for molecule 1 (*solid line*) and molecule 2 (*dashed line*) with deposition points indicated by *arrows*. (From ref. [73])

The dendronized polymers can be considered as linear monodisperse macro-molecule that bear pendant dendrons along the backbone which are recognized as an important alternative structure [84, 85]. These are novel macromolecules whose nanoscale size, rigidity and functionality can be controlled with precision by tuning molecular arquitecture [86]. Nevertheless, dendronized polymer or side-chain dendritic polymers has not received much attention [85, 87–90]. These dendritic macromolecules are characterized by a central polyfunctional core, branching units and end groups. From the core arise successive layers of monomers units with branching points in each monomer unit. This result in a chemical structure that can adopt a spherical shape and where the periphery consists of a large number of chain ends. The side chain dendrons are synthesized with a series of controlled reactions, where each step (generation) results in an exponential increase in the number of monomers. Because of their structural precision they can be considered as synthetic analogue to proteins, [91–93] and there is interest in developing applications in medicine, [94–97] surface science, [98, 99] and catalysis [100]. These applications typically arise from utilizing the large number of functional groups on the periphery, the overall charge of dendritic structure, or property differences between the interior and exterior of the dendrimer [101–104]. Dendritic macromolecules are, in general, represented by two main classes of compound: hyperbranched polymers and dendrimers [105]. Hyperbranched polymers are formed by random or quasi – regular branching of macromolecular segments, as it can be seen in Fig. 4.12a. Dendrimers possess well – defined repeat unit (branches) growing from a central core and forming spherical macromolecules with diameters up to tens of nanometers (Figs. 4.12b, c). Dendritic macromolecules come in thousands of molecular designs depending on the chemical structure and symmetry of branches and the synthetic route applied [105].

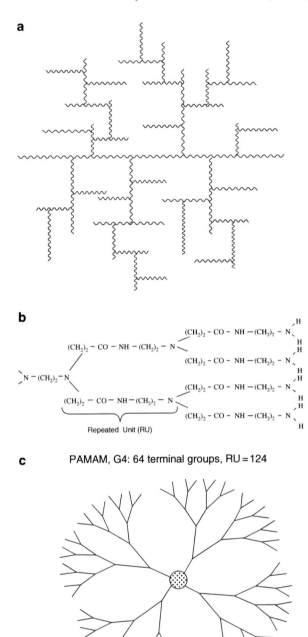

Fig. 4.12 Examples of different dendritic architectures (**a**) hyperbranched macromolecule, (**b**) chemical structure of classic PAMAM dendrimer, repeat unit, and one full branch of generation GI; (**c**) schematic of a backbone for generation G4. (From ref. [106])

Some the structural state of dendritic macromolecules at air – water (Langmuir monolayers) and air – solid (adsorbed monolayers, self – assembled films, and cast films) interfaces has been discussed by V. V. Tsukruk [106] and Frechet [107].

Langmuir mono/bilayers of mono/multidendrons based on 3, 5 – dihydroxybenzyl alcohol at the air – water interface were studied using pressure – area isotherms and neutron reflectivity by Frechet et al. [108].

For asymmetric monodendrons, an increase in compression resulted in collapse of the dendrimer and the formation of a bilayer structure with the macromolecules compressed laterally with respect to the surface normal, with an axial ratio of 2:1. This situation is illustrated in Fig. 4.13. These dendritic macromolecules are flexible enough to assume a prolate conformation under modest lateral compression in a Langmuir trough. Polyamidoamine (PAMAM) dendrimers [109] were also used by Tsukruk et al. [110] to fabricate self – assembled monolayers. The thickness of all the monolayers were much smaller than the diameter of dendritic macromolecules in solution. This behavior can indicate collapse of the dendritic macromolecules, which become highly compressed along the surface normal and flattened. (See Fig. 4.13b).

Watanabe et al. [111] have used an electrostatic layer – by – layer deposition technique to fabricate self – assembled films from alternating molecular layers of oppositely charged PAMAM dendrimers and low molar mass compounds.

They have observed linear growth of the film thickness, which is consistent with multilayer ordering. The thickness of an individual molecular layer for the generation G4 is about 5 nm, which indicates preservation of the globular shape of the dendritic macromolecules within the multilayer films (Fig. 4.13c).

Figure 4.13 shows a general scheme of dendritic macromolecules within molecular layers at interfaces under different situations.

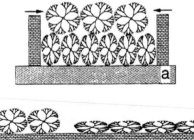

Fig. 4.13 General scheme of dendritic macromolecules within molecular layers at interfaces: (**a**) compressed Langmuir bilayer at air-water interface; (**b**) adsorbed and self-assembled monolayers of "neutral" dendrimers (*left*) and dendrimers with "sticky" surface groups (*right*); (**c**) multilayer self-assembled films obtained by layer-by-layer deposition of dendrimers low molar mass ions (*left*) and two adjacent dendrimer generations (*right*). (From ref. [106])

Several work are concerned with the synthesis, characterization, dielectric behavior and conformational analysis of dendronized Polymers. Poly(methacrylates) containing phtalimidoalkyl moieties in the side chain have been recently studied i.e. poly(3,5-diphtalimido alkylphenyl methacrylate)s with ethyl (P-EthylG$_1$), propyl (P-PropylG$_1$) and butyl (P-ButylG$_1$) chains as spacer groups. Where G$_1$ indicates first generation [112].

In this investigation the dentrimer end groups modifications of a series of PEO-PAMAM linear-dendritic diblock copolymers to various chemical functionalities has been described [112]. The molecular weight of the PEO block was 2000 and the dendrimer generations used were one to four. The amphiphilic behavior of the modified diblocks was studied by spreading monolayers of the material at the air-water interface of a Langmuir trough and recording the pressure-area isotherm.

Stearate-terminated diblocks were found to give stable monolayers which formed condensed phases on compression. The limiting area per molecule in the condensed phase measured from the pressure-area isotherm suggests interesting effect of dendrimer morphology, curvature, and size on the organization of the diblock monolayer at the air-water interface.

The effect of dendrimer generation on the π-A isotherms of stearate-terminated PEO-PAMAM diblock copolymers is shown in Fig. 4.14.

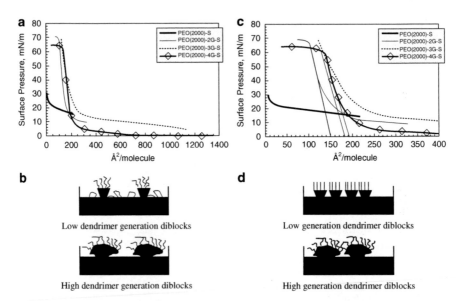

Fig. 4.14 (**a**) Pressure-area isotherms measured at 20°C for four diblock copolymers with stearate end groups and the same PEO chain but different dendrimer generations. (**b**) Schematic of the organization of linear-dendritic diblock copolymers at the air-water interface for different dendrimer generations at high areas per molecule. (**c**) A close-up to condensed-phase regime in the pressure-area isotherms measured at 20°C for four diblock copolymers with stearate end groups and the same PEO chain but different dendrimer generations. (**d**) Schematic of the organization of linear-dendritic diblock copolymers at the air-water interface for different dendrimer generations in the condensed phase. (From ref. [112])

Figure 4.14a shows the π-A isotherm for four polymers, all with the same length of the PEO chain but with different generations of stearate-terminated dendrimers. Again, the area per molecule is calculated from the theoretically expected molecular weight for the diblocks listed in Table 4.4. Two facts area apparent from the comparison shown in Fig. 4.14a. First, the three hybrid linear-dendritic diblocks with stearate end groups form condensed phases at the air-water interface as indicated by the high surface pressures achieved in the isotherm before collapse. Second, the surface pressure for PEO(2k)-4.0G-S at low surface concentration is practically zero, in contrast to all the other polymers investigated here. As mentioned in the previous section, the nonzero surface pressure at low surface concentration is a likely consequence of the surface activity of the PEO block. In the case of PEO(2k)-4.0G-S, the PEO block is being excluded from the interface. A schematic of what could be happening at low surface pressures is shown in Fig. 4.14b. For the hybrid copolymers up to the third dendrimer generation, the PEO tail is long enough to go around the dendrimer block to access the interface at low surface pressure. The dendrimer block in the fourth-generation diblock [PEO(2k)-4.0G-S] is probably larger than the PEO(2k) hydrodynamic radius, and is thus too large for the PEO chain to wrap around and access the interface (Fig. 4.14b).

To take a closer look at the area per molecule in the condensed phase of the diblock monolayers, the x-axis of Fig. 4.14a is expanded and shown in Fig. 4.14c. The extrapolated value of the area per molecule and the theoretically expected value for the area are listed in Table 4.5. To calculate the theoretically expected area for the hybrid block copolymers in the condensed phase, it is assumed that the stearate end groups are extended into the air perpendicular to the air-water interface as shown schematically in Fig. 4.14d. This assumption is based on the behavior of pure stearic acid, which forms ordered monolayers with the alkyl chains oriented perpendicular to the air-water interface. The area per molecule for stearic acid with this orientation is 20 Å^2, [113] Here, the theoretically expected area was calculated by multiplying the area per stearate molecule (20 Å^2) with the number of stearate groups present at the ends of the dendrimer block. PEO(2k)-S having no dendrimer block but a single

Table 4.4 Percentage of Substituted Amine Groups on the Dendrimer Block Determined using ^{1}HNMR and Tabulated as Ratio of the Methyl Groups on the Stearated Ends of the Dendrimer Block to the PEO Backbone Protons

$CH_3O–(CH_2CH_2O)_m–PAMAM–[NHCO(CH_2)_{16}CH_3]_n$

PEO–PAMAM diblock	theor M_w	no end groups (n)	modified diblock	theor M_w	theor ratio (b/a)	exptl ratio (b/a)
PEO(2k)	2000	1	PEO(2k)–S	2267		
PEO(2k)–1.0G	2230	2	PEO(2k)–1.0G-S	2762	0.033	0.033
PEO(2k)–2.0G	2686	4	PEO(2k)–2.0G-S	3750	0.066	0.070
PEO(2k)–3.0G	3598	8	PEO(2k)–3.0G-S	5726	0.132	0.102
			PEO(2k)–3.0G-M^a	4878	0.088	0.087
PEO(2k)–4.0G	5422	16	PEO(2k)–4.0G-S	9678	0.264	0.208

$^a CH_3O–(CH_2CH_2O)_m–PAMAM–[NHCOC_6H_4 \; CH_2 \; CH=CH_2]_n$

Table 4.5 Limiting area per molecule from extrapolation of the condensed-phase region of the isotherms for the modified linear-dendritic diblock copolymers. (From ref. [112])

Diblock	No. stearate groups (n)	Theor area $(20 \text{ Å}^2 \times n)$, Å^2	Exptl area, Å^2
PEO(2k)–S	1	20	25
PEO(2k)–2.0G–S	4	80	150
PEO(2k)–3.0G–S	8	160	185
PEO(2k)–4.0G–S	16	320	195

stearate end groups was studied as a standard and the experimental results agrees well with the theoretically predicted value as it is possible to observe in Table 4.5

Table 4.5 summarizes the limiting area per molecule form extrapolation of the condensed-phase region of the isotherms for the modified linear-dendritic diblock copolymers. As is shown in Table 4.5, the experimental value for the area is larger than the calculated value, suggesting that the condensed phase in this copolymer may contain some PEO at the interface along with the dendrimer block.

The surface properties of this kind of "supramolecular systems" are really scarce.

An interplay between short – range van der Waals forces, ionic binding, chemical bonding, elastic/plastic compression, and long – range electrostatic interactions and capillary forces between macromolecules and surfaces seems to be responsible for the variety of observed interfacial behaviors.

References

1. J. M. Lehn, Angew. Chem. Int. Ed. Engl. 29, 1304 (1990)
2. H. Ogino, J. Am. Chem. Soc. 103, 1303 (1981)
3. G. Agam, A. Zilkha, J. Am. Chem. Soc. 98, 5214 (1976)
4. P. R. Ashton, et al. Angew. Chem. Int. Ed. Engl. 30, 1042 (1991)
5. J. S. Manka, D. S. Lawrence, J. Am. Chem. Soc. 112, 2440 (1990)
6. T. V. S. Rao, D. S. Lawrence, J. Am. Chem. Soc. 112, 3614 (1990)
7. G. Wenz, Angew. Chem., Int. Ed. Engl. 33, 803 (1994)
8. D. Philp, J. F. Stoddart, Angrew. Chem. Int. Ed. Engl. 35, 1155 (1996)
9. A. Harada, Coord. Chem. Rev. 148, 115–133 (1996)
10. W. Hermann, B. Keller, G. Wenz, Macromolecules, 30, 4966 (1997)
11. A. Harada, M. Okada, Y. Kawaguchi, M. Kamachi, Polym. Adv. Technol. 10, 3 (1999)
12. J. Szejtli, Pure Appl. Chem. 76, 1825 (2004)
13. D. Bonacchi, A. Caneschi, D. Dorignac, A. Falqui, D. Gatteschi, D. Rovai, Chem. Matter, 16, 2016 (2004)
14. A. Harada, M. Kamachi, Macromolecules, 23, 2821 (1990)
15. J. Kawasaki, D. Satou, T. Takagaki, T. Nemoto, A. Kawaguchi, Polymer, 48, 1127 (2007)
16. A. Harada, J. Li, M. Kamachi, Nature, 356, 325 (1992)
17. A. Harada, J. Li, M. Kamachi, Macromolecules, 27, 4538 (1994)
18. A. Harada, M. Okada , M. Kamachi, Acta Polym. 46, 453 (1995)
19. A. Harada, T. Nishiyama, Y. Kawaguchi, M. Okada, M. Kamachi, Macromolecules, 30, 7115 (1997)
20. H. Okumura, M. Okada, Y. Kawaguchi, A. Harada, Macromolecules, 33, 4297 (2000)
21. A. Harada, K. Kataoka, Prog. Polym. Sci. 41, 949 (2006)

22. R. Auzely – Velty, M. Rinaudo, Macromolecules, 35, 7955 (2002)
23. F. Djedaini, Ph D. Dissertation, University of Paris X1, France (1991)
24. M.R. Eitink, M.L. Andy, K. Bystrom, H.D. Perlmutter, D.S. Kristol, J. Am. Chem. Soc., 111, 6765 (1989)
25. L. Liu, Q-X Guo, J. Incl. Phen. Macrocyclic Chem., 42, 1 (2002)
26. C.A. Dreiss, T. Cosgrove, F.N. Newby, E. Sabadini, Langmuir, 20, 9124 (2004)
27. T. Ikeda, E. Hirota, T. Aoya, N. Yui, Langmuir, 17, 234 (2001)
28. L. Gargallo, D. Vargas, A. Leiva, D. Radić, J. Colloid Inter. Sci., 301, 607 (2006)
29. T.J. Okubo, J. Colloid Interf. Sci., 125, 387 (1988)
30. B. Szyszkowski, Phys. Chem., 64 385 (1909)
31. J.J. Spitzer, Can. J. Chem., 62 (11) 2359 (1984)
32. K. Yoshida, T. Shimomura, K. Ito, R. Hayakawa, Langmuir, 15, 910 (1999)
33. C. T. O'Konski, "Molecular Electrooptics: Part 1", Marcel Dekker: New York, (1976)
34. T. Shimomura, H. Sato, H. Furusawa, Y. Kimura, H. Okumoto, K. Ito, R. Hayakawa, S. Hotta, Phys. Rev. Lett. 72, 2073 (1994)
35. A. Harada, J. Li, M. Kamachi, Macromolecules, 26, 5698 (1993)
36. A. Harada, J. Li, M. Kamachi, Nature, 370, 126 (1994)
37. I.G. Panova, V.I. Gerasimov, F.A. Kalashnikov, I.N. Topchieva, Polym. Sci., Ser. B 40, 415 (1998)
38. A. Harada, M. Kamachi, J. Chem. Soc., Chem. Commun., 1322 (1990)
39. A. Harada, M. Okada, J. Li, M. Kamachi, Macromolecules, 28, 8406 (1995)
40. H. Jiao, S.H. Goh and S. Valiyaveettil, Macromolecules, 34, 8138 (2001)
41. A. Harada, S. Suzuki, M. Okada, M. Kamachi, Macromolecules, 29, 5611 (1996)
42. Y. Kawaguchi, T. Nishiyama, M. Okada, M. Kamachi, A. Harada, Macromolecules, 33, 4472 (2000)
43. J. Li, D. Yan, Macromolecules, 34, 1542 (2001)
44. J. Lu, I.D. Shin, S. Nojima, A.E. Tonelli, Polymer, 41, 5871 (2000)
45. L. Huang, E. Allen, A.E. Tonelli, Polymer, 40, 3211 (1999)
46. M. Ceccato, P. Lo Nostro, P. Baglioni, Langmuir, 13, 2436 (1997)
47. A. Harada, M. Okada, M. Kamachi, Bull. Chem. Soc. Jpn., 71, 535 (1998)
48. C.C. Rusa, A.E. Tonelli, Macromolecules, 33, 5321 (2000)
49. M. Wei, A.E. Tonelli, Macromolecules, 34, 4061 (2001)
50. C.C. Rusa, T. Uyar, M. Rusa, M.A. Hunt, X. Wang, A.E. Tonelli, J. Polym. Sci. Part A Polym. Phys., 42, 4182 (2004)
51. J. Liu, H.R. Sondjaja, K.C. Tam, Langmuir, 23, 5106 (2007)
52. K.M. Huh, T. Ooya, W.K. Lee, S. Sasaki, I.C. Kwon, S.Y. Jeong, N. Yui, Macromolecules, 34, 8657 (2001)
53. T.E. Girardeau, J. Leisen, H.W. Beckham, Macromol. Chem. Phys., 206, 998 (2005)
54. H. Fugita, T. Ooya, N. Yui, Macromolecules, 32, 2534 (1999)
55. K. Ito, T. Shimomura, Y. Okumura, Macromol. Symp., 201, 103 (2003)
56. T.E. Girardeau, T. Zhao, J. Leisen, H.W. Beckham, D. Bucknall, Macromolecules, 38, 2261 (2005)
57. A. Colombo, G. Allegra, Macromolecules, 4, 579 (1971)
58. C. Choi, D.D. Davis, A.E. Tonelli, Macromolecules, 26, 1468 (1993)
59. A.E. Tonelli, Makromol.Chem. Symp. Ser. 65, 133 (1993)
60. A.E. Tonelli, Macromolecules, 24, 3069 (1991)
61. A.E. Tonelli, Macromolecules, 24, 1275 (1991)
62. H. Ritter, M. Tabatabai, Prog. Polym. Sci., 27, 1713 (2002)
63. V. Alupei, H. Ritter, Macromol. Rapid Commun., 22, 1349 (2001)
64. L. Lethen, J-M. Kim, D.H. Thompson, Polym. Rev., 47, 383 (2007)
65. H. Frey, Angew. Chem. Int. Ed. 37, 16, 2193 (1998)
66. S. Grayson, J.M. Frechet, Chem. Rev., 101, 3819 (2001)
67. H. Frauenrath, Prog. Polym. Sci., 30, 325 (2005)
68. R. Tang, Z. Tan, C. Cheng, Y. Li, F. Xi, Polymer, 46, (14), 5341 (2005)

69. M. Calderon, M. Martnelli, P. Froimowicz, A. Leiva, L. Gargallo, D. Radić, M.C. Strumia, Macromol. Symp., 258, 53 (2007)
70. Y. Kim, J. Pyun, J.M.J. Frechet, C.J. Hawker, C.W. Frank, Langmuir, 21, 10444 (2005)
71. Y. Seo, A.R. Esker, D. Sohn, H.J. Kim, S. Paek, H. Yu, Langmuir, 19, 3313 (2003)
72. M. Lee, C.J. Jang, J.H. Ryu, J. Am. Chem. Soc., 126, 8082 (2004)
73. J. Holzmueller, K.L. Genson, Y. Park, Y-S. Yoo, M.H. Park, M. Lee, V. Tsukruk, Langmuir, 21, 6392 (2005)
74. Q. Ma, E. Remsen, C.G. Clark, T. Kowalewski, K.L. Wooley, Proc. Natl. Acad. Sci., 99, 5058 (2002)
75. J.A. Hubbell, Science, 300, 595 (2003)
76. D.J. Pochan, Z. Chen, H. Cui, K. Hales, K. Oi, K.L. Wooley, Science, 306, 94 (2004)
77. G. Riess. Prog. Polym. Sci., 28, 1107 (2003)
78. I.W. Hamley, "The Physics of Block Copolymers", Oxford University Press, Oxford (1998)
79. D.E. Discher, A. Eisenberg, Science, 295, 967 (2002)
80. M. Maskos. Polymer, 47, 1172 (2006)
81. D. Yan, Y. Zhou, J. Hou, Science, 303, 65 (2004)
82. R. Mezzenga, J. Ruokolainen, G.H. Fredrickson, E.J. Kramer, D. Moses, A. Heeger, O. Ikkala, Science, 299, 1872 (2003)
83. T.P. Russell, Science, 297, 964 (2002)
84. R. Yin, Y. Zhu, D.A. Tomalia, H. Ibuki, J. Am. Chem. Soc., 120, 2678 (1998)
85. M.G. Scott, J.M.J. Fréchet, Macromolecules, 34, 6542 (2001)
86. Ch. Zhang, L.M Price, W.H. Daly, Biomacromolecules, 7, 139 (2006)
87. J. Das, M. Yoshida, Z.M. Fresco, T-L. Choi, J.M.J. Fréchet, A.K. Chakraborty, J. Phys. Chem., 108, 6535 (2005)
88. A. Zhang, L. Shu, Z. Bo, A.D. Schluter, Macromol. Chem. Phys., 204, 328 (2003)
89. C. Ecker, N. Severin, A.D. Schluter, J.P. Rabe, Macromolecules, 37, 2484 (2004)
90. L. Shu, A. Schluter, C. Ecker, N. Severin, J.P. Rabe, Angew. Chem. Int. Ed., 40, 4666 (2001)
91. O.A. Matthews, A.N. Shipway, J.F. Stoddart, Prog. Polym. Sci. 23, 1 (1998)
92. C.R. DeMattei, B. Huang, D.A. Tomalia, Nano Lett., 4, 771 (2004)
93. N.W. Suek, M.H. Lamm, Macromolecules, 39, 4247 (2006)
94. U. Boas, P.M.H. Heegaard, Chem. Soc. Rev., 33, 43 (2004)
95. A.J. Khopade, F. Caruso, Biomacromolecules, 3, 1154 (2002)
96. S. Fuchs, T. Kapp, H. Otto, T. Schoneberg, P. Frank, R. Gust, A.D.Schutler, Chem. Eur. J., 10, 1167 (2004)
97. S. Chauhan, N.K Jain, P.V. Diwan, A.J Khopade, Drug Target, 12, 575 (2004)
98. V.V. Tsukruk, F. Rinderspacher, V.N. Bliznyuk, Langmuir, 13, 2171 (1997)
99. A.W. Bossman, H.M. Janssen, E.W. Meijer, Chem. Rev., 99, 1665 (1999)
100. R.M. Crook, M. Zhao, V. Chechik, L.K. Yeung, Acc. Chem. Res., 34, 181 (2001)
101. A. Quintana, E. Raczka, L. Piehler, I. Lee, I. Majaros, A.K. Patri, T. Thomas, J. Mule, J.R. Baker, J. Pharm. Res., 19, 1310 (2002)
102. M. Liu, K. Kono, M.J. Fréchet, J. Polym. Sci. Polym. Chem., 37, 3492 (1999)
103. K. Kono, M. Liu, M.J. Fréchet, Bioconjugate Chem., 10, 1115 (1999)
104. H. Kaya, N-R. de Souza, J. Appl. Cryst., 37, 223 (2004)
105. G.R. Newkome, C.N. Moorefield, F. Vogtle (Eds), "Dendritic Molecules", VCH, Weinheim (1996)
106. V.V. Tsukruk, Adv. Mater., 10, 3, 253 (1998)
107. J.M.J. Frechet, Science, 263, 1711 (1994)
108. J.M.J. Frechet, C.J. Hawker, K.L. Wooley, Pure Appl. Chem., A31, 1627 (1994)
109. D.A. Tomalia, A.M. Naylor, W.A. Goddard, Angew. Chem. Int. Ed. Engl. 29, 138 (1990)
110. V.V. Tsukruk, F. Rinderspacher, V.N. Bliznyuk, Langmuir, 13, 2171 (1997)
111. S. Watanabe, S.L. Regen, J. Am. Chem. Soc., 116, 8855 (1994)
112. J. Iyer, P.T. Hammond, Langmuir, 15, 1299 (1999)
113. D. Myers, "Surfaces, Interfaces and Colloids, Principles and Applications", VCH Publishers: New York, 165 (1991)

Index